On Science
On Art
On Society

Interviews
with Innovators

EDITORS

Arthur Clay, Monika Rut, Timothy J. Senior

Published, sold and distributed by:
River Publishers
Niels Jernes Vej 10
9220 Aalborg Ø
Denmark

River Publishers
Lange Geer 44
2611 PW Delft
The Netherlands

Tel.: +45369953197
www.riverpublishers.com

ISBN: 978-87-93379-12-1 (Paperback)
      978-87-93379-13-8 (Ebook)

©2015 River Publishers

"The imagination exercises a powerful influence over every act of sense, thought, reason, over every idea."

Latin Proverb

Contents

On Science On Arts On Society

An Introduction by the Editors 1

Introduction 1

The Origins of On Science, On Art, On Society 2

Acknowledgments 3

The Beauty of Life: Art and Science

An Interview with Sunghoon Kim 7

DIYbio & Democratizing Science

An Interview with Denisa Kera 15

Where Art & Science Coexist

An Interview with Catherine Young 21

Atlas Remeshed

An Interview with Davide Angheleddu 28

A Full Understanding of Things

An Interview with Ruedi Stoop 38

Neuroaesthetics: Beyond Mere Beauty

An Interview with Jaeseung Jeong 46

No Science without Humor

An Interview with Dr. Anatophil 52

Art, Science, and The Brain

An Interview with K-soul 58

Dubbing and Dubbing It All

An Interview with Simone Carena 67

Augmented Increments

An Interview with Felix Heisel 76

The Museum and You

An Interview with Boa Rhee 81

Keeping It Healthy and Curious

An Interview with The Curious Minded 88

Curating Infinity

An Interview with Houngcheol Choi 93

We Are All the Dust of Stars

An Interview with Choi Jeong-hwa 101

The Harmony Amongst Us

An Interview with Hyung Joon Won 105

Connecting Art and Science

An Interview with Jill Scott 112

Creative Collisions: Arts and Science

An Interview with Ariane Koek 120

Winning Opportunities

An Interview with Raphael H Cohen 128

Approaches to Responsible Behavior

An Interview with Susan Schneider 134

Driving the World with Social Innovation

An Interview with Byungtae Lee 141

Making Impact First

An Interview with Chong Soo Lee 151

Hierarchy of Failure, and Solutions

An Interview with Raymond Saner 157

Balancing the Planet with Play

An Interview with Sun Mi Seo 166

Learning By Gaming

An Interview with Peter Lee 173

Benchmarking Eco-Friendly Design

An Interview with Jun Seo Lee 180

From Collisions to Collaborations

An Afterword by Jeungmin Noe 188

Glossary   **189**

About the Editors   **193**

On Science On Arts On Society

An Introduction by the Editors

Introduction

The book "On Science, On Art, On Society: Interviews with Innovators" features a collection of twenty-five interviews with practitioners who are testing the boundaries between the sciences, arts and humanities. Divided into three chapters – On Science, On Art, and On Society – these interviews illustrate to the reader how models of hybridity can nurture new forms of innovation across disciplines and sectors.

Fostering innovation in this way is not a simple linear process, but, rather, a complex and multifarious one. It requires the networking of different resources and talents, the pooling of skills, the challenging of boundaries, and the exploration of risk. In facing many of our twenty-first century problems, new forms of convergence between disciplinary practices are now seen as a necessity, and point to an understanding of innovation that must include some form of hybridity if it is to prove effective.

Models of hybrid innovation that can draw on diverse resources, and which can be implemented as a catalyst to trigger innovation, are becoming increasingly common in a variety of settings. A sense of urgency can be felt amongst many communities to reach outside the four walls of their institutions and establish – with all force – innovative platforms that can help tackle the complex challenges and opportunities they encounter. Amongst the many topics addressed in this book, for example, authors ask whether neuroscience can uncover the laws of aesthetics, whether citizen science will open scientific research to a larger community of stakeholders, whether corporate responsibility and grass-roots innovation can bring about a fairer society based on the sharing of opportunities and resources, and whether new approaches to tourism can help build inclusive and sustainable local economies.

By providing platforms for knowledge exchange and co-creation across different disciplines, we will see the continued emergence of new forms of understanding and insight. Taken together, we are seeing the potential for hybrid models of innovation to usher in an era of more sustainable and responsible governance, economics, research, education and citizenship.

Readers interested in innovation, and the processes that drive it, will find in each themed chapter unique insights from established, as well as upcoming, innovators. Offering examples from their own fields of expertise, we hope these interviews will act as a spark for future discussion, and inspire others to explore the exciting possibilities that cross-disciplinary collaboration can offer. We also hope that the Korean-Swiss context from which the book stems may serve as a model for nurturing cross-cultural exchange in this endeavor.

The Origins of On Science, On Art, On Society

The idea of creating a book of interviews came about during the eighth Digital Art Weeks (DAW) Festival held at the Museum of Art, Seoul National University, in the fall of 2014. Founded and directed by Arthur Clay, DAW International organizes programs that consist of symposia, workshops, exhibitions, and performances exploring the application of digital technology in the arts. The festival's aim is to make artists aware of emerging technologies in the sciences, and scientists aware of how such technologies might be applied in the arts. The DAW Seoul 2014 program focused on the possibilities of convergence between the arts and the sciences within the socio-cultural context of South Korea, drawing together a number of internationally established research institutes as well as upcoming entrepreneurs, artists and thinkers. What began as the "Hybrid Highlights – Switzerland and Korea" exhibition and three Innovation Fora on the theme of convergence (called On Science, On Art, and On Society) now concludes here in this book of interviews with those who took part in the festival, innovators whose insight and expertise make this work possible.

Acknowledgments

Our special thanks goes to Juerg Gutknecht (ETHZ), Juerg Brunnschweiler (ETHZ), Gerhard Schmitt (FCL SEC), Sunghoon Kim (SNU), Simone Carena (Hongik), Lichia Yiu Saner (CSEND), Raymond Saner (CSEND), Jeugnmin Noe (SNU MoA), Sean Hill (EPFL), Jusub Kim (Sogang) who helped shape the design of the Innovation Forum and offered guidance throughout its planning. We would also like to thank DAW sponsors who generously gave their support: Presence Switzerland, ETH Zurich, Swiss Arts Council Pro Helvetia, Embassy of Switzerland in the Republic of Korea, Swiss Cultural Fund Korea, Swiss State Secretariat for Education, Research and Innovation (SERI), and Seoul National University along with its adjunct Museum of Art Seoul National University (SNU-MoA).

ON SCIENCE

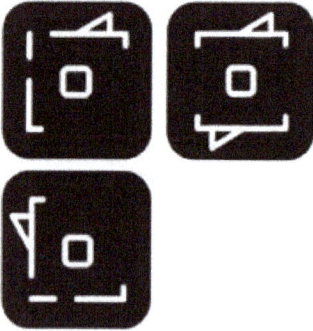

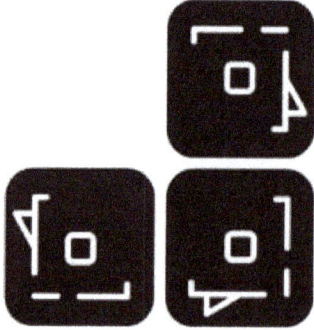

The pervasive nature of science today is bringing it into contact with a greater range of disciplines and practices than ever before. Science, as we know it, is changing through the tangential influences it exerts and the unexpected reactions it triggers in return. The question of what "Science" now constitutes demands attention. On the one hand, improved access to the ideas, methods and products of science is opening up new forms of interface with the public and those communities affected by its research. On the other hand, science practitioners are testing where the limits, as well as new prospects, lie in their work, breaching disciplinary boundaries to spark debate and controversy. Here, we have the opportunity to reflect on the wider repercussions of our scientific expansion, its social and cultural responsibilities, and to ask where participation in science begins and ends.

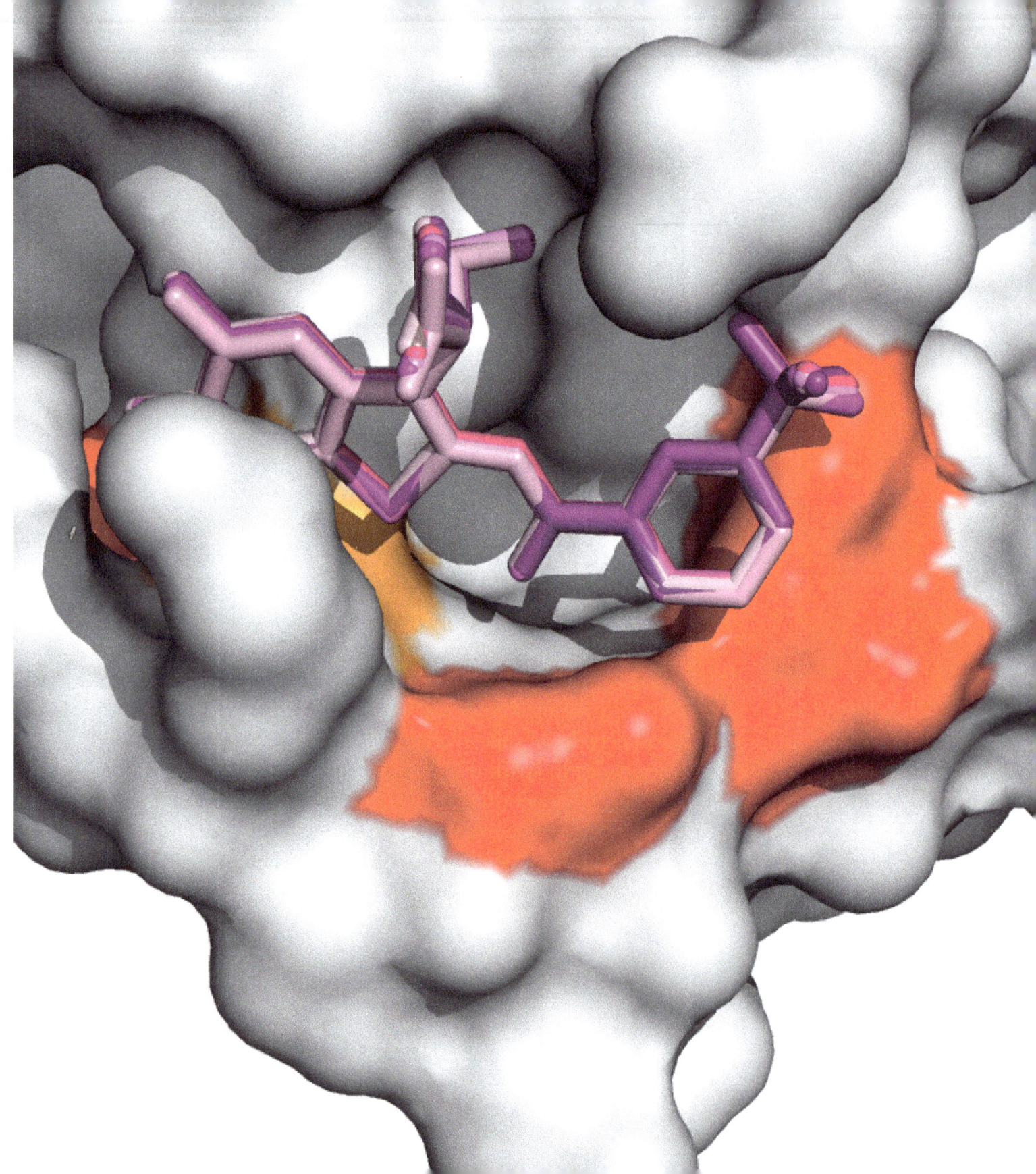

The Beauty of Life: Art and Science

An Interview with Sunghoon Kim

Interviews with Innovators: You are the director of the Medicinal Bioconvergence Research Center (Biocon), which aims to identify new markers of human disease that can stimulate therapeutic-drug development. Can you tell us briefly about Biocon and its goals?

Sunghoon Kim: Biocon was established in 2010 under the Global Frontier Program initiated by the Ministry of Science, ICT and Future Planning, South Korea. Its founding vision was to drive innovation in drug discovery through the application of novel science and convergence technologies. Currently, drug development is suffering from high failure rates and high costs, and the barrier to new drug development is getting higher every year. This is not only a problem in South Korea, but a global issue as well. At the time, the South Korean government felt urgently that they had to do something to ease through the barriers to drug discovery and help this country's relatively small pharmaceutical industry. Hence, Biocon was created with a primary role of helping the drug industry to conduct breakthrough science and drive the development of innovative technologies in this area. In its short-term goals, Biocon strives to identify novel drug targets, carry out drug development, and develop innovative instruments and devices that can generate new drug leads in smarter, more accurate, faster and economically viable ways (we call this technology code S.A.F.E.). In the long term, Biocon is trying to become a world leader in the identification of suitable targets for innovative drug development, to become a world-leading "Target factory".

IWI: There is often talk about twenty-first century epidemics arising through new pathogens, increasing drug resistance, larger and more mobile human populations, and so on. What are some of the key challenges faced by the pharmaceutical industry in its search to create drugs that are effective, available and affordable?

SK: A big conflict in the drug industry is that companies are reluctant to develop drugs against diseases killing people in poorer countries for reasons of profit. They only show an interest when such diseases begin to affect wealthy countries, as we have seen in the recent case of Ebola in parts of Africa. To tackle this problem, and to make drugs accessible and affordable to poor countries, we need another approach to developing drugs on a nonprofit basis for the general health and wellbeing of human society. This will continue to be a major challenge in the future.

IWI: In addition to your work with Biocon, you are an integral part of the Yadahaus, a cultural center in Doguk that offers a space to view art and share in cultural activities. How did Yadahaus come about, and what does it aim to achieve?

SK: Yadah is a Hebrew verb meaning "worship or songs of praise". My wife and I are passing through middle age, living as artist and scientist respectively (and both professors). Looking back over our lives so far, we feel appreciation more than pride. We feel that we have gained more than that achieved through our efforts alone; even this house was inherited after my father passed away. Therefore, we would like to share what we have achieved, to share a gift with other people and society in some way. The Yadahaus was founded on this idea: we want it to be used as a space for art, conversation and sharing. The place is designed as a "melting pot" of peoples regardless of their background and age group.

IWI: Through these two sides to your work – the scientific and the cultural – we see a concern not only for pushing boundaries in innovation but opening up discussion about the world we live in. How important is it, as a state-funded company, to open channels of communication with the public to discuss Biocon's work?

SK: This could be a question posed to many scientists around the world since the majority of scientists – those in the public domain – are only able to fund their research because of taxpayers' money. It is a real challenge for these scientists to convince society why either government or non-scientists should support their research when much of it is conducted for curiosity alone and without any clear, immediate real-world application. Although the scientific method originated with intellectual curiosity, today it is more closely related to industry and wealth creation. In this connection emerges a clearer justification for why the public sector should support the work of scientists. Biocon was established, and is supported, by the South Korean government based on exactly this rationale: its specific goal is to help both the drug industry and improve human health. The question now becomes how to convince the public that Biocon should receive the support that it does.

Although Biocon is quite visible in the public domain, the discoveries we make that could be of interest to a wider public are often only barely, or briefly, recognized by media channels. I think scientists need to find more effective and interactive ways of communicating their work to people outside of their research field. This was the principal reason why I became interested in linking the life sciences to arts and culture, so that non-scientists might gain further interest in our work and become supporters of scientific research. I think those events in which both scientists and the general public can come together to talk about research and its consequences are much more effective – or certainly a strong alternative – to communicating with the public through a jargon-laden press release every once in a while.

IWI: Biocon has initiated several out-of-house activities to facilitate this pubic communication, including a "bio-art" competition and the curation of science-themed artworks for new exhibitions. Who are you targeting with these activities, and what do you hope to achieve?

SK: One example of our work in this area is the cartoons we had commissioned to explore some of the research stories Biocon have published. Cartoons can be a very powerful way of extracting the essence of a given piece of research and deliver these discoveries to a general audience in a more accessible and enjoyable way. These activities have two target groups expressly in mind: the general public outside of science, and the scientific community itself. For non-scientists I am trying to deliver two key messages: The first is to share with them the beauty of life in a new way; the second is to make them more familiar with the work and research interests of the scientific community. The second group consists of scientists themselves, those who have provided the content and ideas that lie behind these works. For scientists, I want to provide them with an opportunity to explore new ideas and perspectives they might not have previously considered. If you are on the search for inspiration, creativity and wonder, you just need to look deeply into life itself; it is just a matter of whether you have the room in your mind to see it. I hope that, through these efforts of ours, more people will begin to appreciate – and protect – the beauty of life.

IWI: Taking this further, do you see a deeper link between works of fine art inspired by science and scientific visualizations created to better communicate the world of scientific discoveries?

SK: In my eyes, artistic and scientific practices have many things in common, even if they use different tools, methods and media. The reason I like to go deep into my research is because I feel I can only see the beauty of nature when I get closer to its "real truth". Since nature is a fine art in its own right, all we should aim to do is expose the original as much as possible. Sometimes non-scientists cannot see the full depth of this beauty in nature without help. (It is like learning to enjoy classical music: it can take time and training before you really appreciate it). I think we often need another approach to help unveil the beauty of nature to the un-trained eye. So, fine art inspired by science and science visualization have much in common in that they both display the beauty of nature. However, I see fine art as the client of nature, whereas scientific visualization is a mediator between nature and the wider public.

IWI: Biocon has signed Memoranda of Understanding with a number of venues including museums, cultural centers and galleries to co-organize its exhibitions with the intention of educating the public about the biosciences. What challenges do you face in this process?

SK: Fostering links between the arts and sciences cannot be done alone: I need support and help from across different communities to make these activities possible. The support we look for varies a great deal but can often include financial assistance, the provision of exhibition spaces and the donation of people's time. Finding a good venue is particularly challenging since most prestigious galleries are still reluctant to display artworks created by amateurs, and most professional artists are not interested in showing their work in non-professional venues. There are, however, galleries and art spaces that share the vision of our work, and will hopefully continue to kindly provide us with space and support for our exhibition projects.

IWI: Are there challenges that scientists will face if they wish to engage with the arts in this way?

SK: The quality of scientific research is normally assessed through its academic and industrial impact, or the number of research publications and patents generated. While academic impact is measured by a range of citation indexes (or similar criteria), industrial impact is judged according to the financial gain (or other forms of tangible value) generated by that particular discovery or invention. Although this type of evaluation has been useful in comparing the relative performance of scientists, it may also restrict the wider flow of ideas both within science and between sectors. This is because any activity outside of this framework won't contribute to the measure of a scientist's success. In this evaluative system, there is no rational ground on which to justify the art-science activity that Biocon has organized in recent years since it contributes neither to basic scientific research nor industrial advancement. So, perhaps we need another way of evaluating performance in science from the point of view of creativity and innovation, even if that activity does not necessarily bring financial gain or new products to the public domain in the short-term.

IWI: Although there are many challenges faced, do you feel that there is much to gain through closer collaboration between artists and scientists?

SK: I believe that engaging with the arts is becoming an increasingly important part of being a good scientist. Not only can such connections be a powerful way of communicating scientific research to a wider audience, but it also exposes scientists to new ways of thinking about their own work. However, unless you are born with a talent for both the arts and sciences, you will need to find artists who show an interest in your work and are willing to collaborate. This is by no means an easy task! I also believe that such collaborations would be of huge benefit to artists. Indeed, I would strongly recommend artists with an interest in the sciences to look into forming collaborations with the scientific community. I think that an exploration of the sciences through the arts has great promise in opening up new artistic territories, both within the fine arts but also in design fields.

Sunghoon Kim is Director of the Medical Bio-convergence Research Center (Biocon) and a professor at the Molecular Medicine and Biopharmaceutical Science Department of Seoul National University. He received his Ph.D. in Biology and Medicine from Brown University. Since 1998 he has served as the director of several research centers including the National Creative Research Initiatives Center for ARS Network, the Information Center for Bio-pharmacological Network, the Integrated Bioscience and Biotechnology Institute, and Biocon. He has also taught graduate programs in bioinformatics, genetic engineering, and cancer biology at the College of Pharmacy, Seoul National University.

DIYbio & Democratizing Science

An Interview with Denisa Kera

Interviews with Innovators: You are a scientist, a philosopher, and a designer with an interest in opening up scientific practice – its ideas and methods – to a wider public. This has led to an involvement with the citizen science network Hackteria. How did this collaboration come about?

Denisa Kera: I am trying to connect my interests in science, philosophy, and design through projects which strongly support open science and citizen science goals. I believe that it is essential to democratize science, especially in the **Global South**, and that is how I met other Hackteria members. We simply share similar values and in some cases even lifestyles. I am something of a nomad with a cause: I travel around the world to visit and work with hacker spaces, Do-It-Yourself biology (DIYbio) labs, and various other forms of grassroots citizen science movements. As a researcher I'm curious to understand what is the best way for a society to interact with, adopt, and integrate emergent technologies. I like these spaces and networks (like Hackteria) because they let everyone understand and take part in the process of designing, tinkering, and playing with various ideas around science and technology. I'm also excited about the prospects of alternative R&D models that can work in developing countries such as Indonesia or Nepal.

IWI: You have been working with the artist and scientist Mark Dusseiller for several years now. When and where did you meet him?

DK: In January 2012 I visited Fablab Yogyakarta in Indonesia, where I met Marc Dusseiller. We were both fascinated by the mobile-food culture in Indonesia, and decided to "hack" one of these mobile food trucks which park in different parts of the city to attract local communities and serve as a "public forum" for people to exchange information and gossip while they share meals. We were convinced that these mobile Warungs or Angkringans are a perfect model for citizen science labs based on interaction with the public.

The "Hacking Angkringan" project transformed one of these cars into a molecular gastronomy lab with basic microbiology equipment. We decided to perform this posh cuisine using scientific principles on the streets of Indonesia, testing and showing how simple equipment such as webcams can be turned into microscopes, and how appropriated scanners can serve as sterilization chambers. For Marc, this approach has been part of many of his past projects: He has often worked with local designers and scientists on open hardware solutions built from cheap lab equipment. For me, at the time, it was more about molecular gastronomy.

IWI: What plans do you currently have within the Hackteria network?

DK: Through Marc I met the rest of Hackteria that year, including Urs Gaudenz (an **open source hardware** prodigy) and Sachiko Hirosue (a scientist from the Swiss Federal Institute of Technology in Lausanne with a passion for open science), and we became inseparable. I visited them later in 2012 after I had my brain scanned in London to perform "brain uploading" over Dropbox and Facebook, a project where I would use the fMRI scan data acquired for a workshop on data liberation as part of a citizen science project in Prague. We spent a lot of time discussing, and quarreling about, similar amateur attempts to understand DNA or fMRI data. This is just to give you some idea of how fluid and collaborative these projects are.

Since then I regularly take part in night hunts for fluorescent organisms in the Swiss Mountains – last time it was using "hacked Indian spiritual chants automata" – and I hope one day Hackteria will meet Brmlab, my Prague hacker-space people, and join together on a hunt for meteorological stations. I love the similarity between these two activities: Basically you try to catch hardware as if it were prey. Based on information that some meteorological station has completely broken down, you locate it and scavenge the hardware to turn it into something different. Essentially, you're turning these trophies you've hunted into open hardware projects; you're liberating them! While I still work with other hacker spaces and groups, my closest friends and connections are people from Hackteria in the EU, Indonesia, and now also in Singapore. The network is spreading and morphing like a living organism.

IWI: "Hackteria" as a term can be understood as bringing together "hacking" and "bacteria" – with a particular emphasis on DIY – to produce interfaces between technology, nature and people. How do Hackteria projects function within diverse environments of artistic, scientific, and community practice?

DK: I would describe most Hackteria (**biohacking**) projects as playful. I find most **Bioart** projects, or interdisciplinary art-science projects, too serious and tedious; they often make science even more complicated and abstruse than it is. I think we are trying to achieve the opposite: to demystify science, to make it part of everyday life and everyday interests such as cooking, hiking or entertaining (especially entertaining kids), to enjoy it fully with friends. Then we will all – across the Global South – be truly committed to science which supports open access, open data and open- and low-cost infrastructure.

We are curious to see how people use and misuse science to their own ends, for their own art projects, to serve their own communities. For this reason I have a problem with many science communication projects, particularly those which are simply trying to perform a PR function and manipulate the public into thinking in a particular way. Through democratizing technology and knowledge, and allowing more horizontal relations between people to be built in a less-intimidating atmosphere than a school, you can then learn something new by playing and tinkering, by teaching other people, by suddenly realizing that what you have learned could serve a particular need faced by your community.

IWI: Open data and open source tools are important to hacker culture. How do hacker spaces, DIYbio and mobile labs make use of open source tools and open data?

DK: Now seriously, hacker spaces and open-source approaches are, so far, the best response to the criticism that the pace of our technological progress doesn't follow either our moral and social progress or our aesthetic sensitivities, the criticism that somehow we are creating things and technologies which do not make us better as individuals or help communities. The typical geek will never argue about this, finding the whole discussion ridiculous: She never knows what it is she is building; she doesn't make a distinction between scientific and social value; she accepts uncertainty and likes to explore possibilities; she trusts the global community of geeks and enthusiasts who will test and take part in her adventure, challenge her and then collectively, tentatively decide on the next steps to take. These spaces are democratic to their core, with innovation constantly evolving through collective experimentation based on tinkering and testing rather than on a big theory of social order or some scientific breakthrough, or a new disruptive technology etc...

Hacker spaces are my playground, or special zone, where I go to dream, where I feel free to test various crazy ideas around emergent technologies. Right now I'm curious about how microfluidics and lab-on-a-chip technologies can be used to make a puppetry show with small organisms, another project with a long Hackteria history. Before this, I was looking into food applications. I'm still an advocate for open biodata and I'm interested in various forms of sharing biodata (fMRI and not only DNA). In another project I was trying to connect open hardware with traditional crafts in Japan. What is great for me as a designer in these hacker spaces is that I get to understand and test how these technologies will work in various cultural and social contexts.

IWI: Hackteria Network is supported largely by public and governmental agencies through diverse types of grants. How does this shape the content

of projects and who participates, and is this approach to receiving sup-
port a viable means of sustainability for the network?

DK: Actually, many of the activities are also self-funded, but it depends. I wish there was more support with no strings attached, but in that sense I have to say I'm also impressed by grants from Migros and Pro Helvetia who were generous to support projects with such a strange combination of science, art, design and geopolitical networks – projects that don't make any sense. I think we are putting Switzerland on the map as a place not only for cutting-edge technology, but also humanitarian R&D efforts which democratize science. These efforts enable more people to enjoy science, take part in it, and, through overcoming fear of it, develop the abilities they need to assess its implications and decide for themselves what sort of work they wish to see in the future.

Denisa Kera is a philosopher and designer with an interest in open science and citizen science. She explores these themes through building prototypes and critical probes that serve as tools for deliberation, reflection and public participation. A core focus of her work is alternative R&D spaces that can serve as models for citizen science, including Hackerspaces, FabLabs, and diverse "Do It Yourself Biology" (DIYbio) movements around the world. Her projects address, for example, issues in food design, the use of DNA and fMRI data, and open software/hardware in science. She has extensive experience as a curator of exhibitions and projects related to art, technology, and science, building on an earlier career in Internet startups and journalism. Currently, she works as an Assistant Professor at the National University of Singapore, where she is also an Asia Research Institute and Tembusu college fellow.

Where Art & Science Coexist

An Interview with Catherine Young

Interviews with Innovators: As an artist, your work primarily explores human perception and its relationship to memory, creativity, and play. In a way, your story as an artist begins with studying molecular biology and biotechnology at the University of the Philippines. What inspired you to move into the arts? What did you take with you from the sciences?

Catherine Young: I come from a family of artists and doctors. Although my grandfather was a photographer, my mom was a genetics professor which meant I grew up with a lot of Punnett Squares and DNA lessons. I've always loved science, and although I initially wanted to do only lab-based research I soon realized that being cooped up in the lab just wasn't for me – there were so many other ways I wanted to pursue my ideas. I moved to New York City at the age of 21 and just saw the endless possibilities that lay ahead of me instead of the single, stifling path in science I had chosen. I moved – or more specifically, "ran away" – to Barcelona and rediscovered art, but also discovered for the first time how art and science could work together. From there I chose to study interaction design for my MFA in the US (which was a relatively new field at the time) because I felt it was a discipline where art and science could coexist and be effectively communicated to others.

IWI: You have worked in a variety of traditional forms including drawing and painting, but also with less-traditional media such as dirt or soil. What roles have artistic and scientific practices played in helping you break away from traditional formats?

CY: When two things collide, new relationships can be formed. For example, visual art is usually composed of traditional fields such as painting or sculpture – works that the public shouldn't touch else you damage them. But this doesn't hold true with interactive pieces where the work has to be touched to be experienced. The convergence of art and science leads to new ways of offering stories for people to explore, and unique opportunities for empathy. Also, advancements in science along with new ways of expression in art are leading to greater diversity in the type of work that can be done. For instance, there's been a relatively recent interest in art investigating smell because of research that shows olfaction is linked to memory. So, if I were an artist interested in memory, this sense is now a channel for me to investigate that theme. Science can inform art and vice versa; each enriches the other.

I don't really look at my projects and say: "Ok, what is the 'art' and what is the 'science' here?" I start with the questions I want to explore, and then test those approaches that I feel will best communicate these ideas I want to share with people. My encounter with different artistic and scientific fields has exposed me to a lot of different ways for investigating topics and then presenting them to a wider audience. Science for me involves a lot of data collection, something which I often do in my current work. Artistic methods help me explore the many ways in which I can then show that work. Each exhibition of a project is also like an experiment to me, so my work continues as I explore how people respond to it. Personally, I don't think there's much of a difference between art and science as both ask similar questions. However, there is a great difference in the professions of art and science because of the systems we have created to practice them, such as the galleries, festivals, labs, academic and industrial institutions, and so on.

IWI: "Disclosure of knowledge" or "lifting of the veil" is how you describe your Apocalypse Project. Here you are trying to physically animate an inquiry made concerning environmental futures. Can you tell us about this research in general and what the process was that transformed this enquiry into an artwork?

CY: I started The Apocalypse Project during my participation at the 2013 ArtScience Residency Program in partnership with the ArtScience Museum at Marina Bay Sands, Tembusu College National University of Singapore, and the Singapore-ETH Zurich Future Cities Laboratory (FCL). I was talking to the scientists in FCL who were doing very interesting research on climate change and sustainability. But, while their work was important, I saw a gap between their research and the public understanding of climate change.

In parallel to this, I was doing workshops with high school and college students in Singapore, asking them questions about what climate change looked like to them, what superpowers they wished they had to combat climate change, and what they would wear to a climate-change apocalypse. The last question resonated most strongly with the majority of the students. I realized that fashion is one thing that people can relate to – clothing is both a means of survival and a form of self-expression. This was the beginning of Climate Change Couture. I took the research of the FCL scientists and designed clothes and narratives that would fit a world that was uninhabitable, highlighting the problems that the research projects addressed. Members of the FCL staff took on the role of models for the first collection. I think it was wonderful to see them outside the lab and wearing these outfits.

IWI: Besides showing the actual objects themselves, how else have you tried to communicate the fact that climate change is real, that "the heat is on" so to speak?

CY: The projects are all about experience. So, rather than just presenting the clothes or the perfumes I've developed, the visitors are invited to try on the clothes themselves and smell the perfumes. In this way, they are able to place themselves into the story. I have also held public events, such as Future Feast at The Mind Museum in Manila, where I got to collaborate with chefs to think about dishes of the future. It was a real feast with local musicians performing, scientists explaining topics about climate change, chefs talking to people about why things like worm meat and sea vegetables could be future sources of nutrition – it included activities for the whole family. I think doing these inclusive and fun events reframes climate change from a doom-and-gloom political issue (meant to be discussed only by governments) to a human issue about creativity and resilience that everyone should act on.

IWI: Some of the garments that you have put on exhibition connect traditional design culture with what we might call "a fashion of necessity". Does the inclusion of traditional elements make the message clearer that climate change and cultural change are synonymous?

CY: I think culture has always had to adapt to the environment. For example, the Barong Tagalog, which is a traditional dress for men in the Philippines, was designed to be lightweight because of the country's tropical climate. I re-imagined it with a hoodie to reflect the unpredictable weather that the country now has. When designing for a particular city, I like to research their traditional garments because the message I want to communicate will resonate better with an audience if the visual imagery is familiar and something they can relate to. A lot of my projects deal with future loss, and so the audience has to imagine a world where some things are not available to satisfy the needs of their traditional practices.

IWI: Which of the approaches used to carry your message have been most effective in motivating and empowering people to become co-creators of a more tangible future?

CY: I think it's the fact that I do multiple projects, so I give people different ways to engage with environmental futures. If one doesn't do it for them, another one might. Each project is also exhibited through multiple platforms, in multiple cities, and uses different types of media. Some have encountered the projects online, but, for many, the work is more powerful when seen and experienced in person. From what I've observed, I think the work becomes most effective when people are able to share their experiences and the memories that these projects evoke with other people. This allows conversation about climate change to reach beyond the exhibition or the festival and into people's normal everyday lives. I also target a wide audience base. I'm particularly interested in the reactions of young children because they are the most honest. They will also bear the worst consequences of climate change, so I think they need to learn about it and be taught as soon as possible how to take care of the environment.

IWI: Knowing that the research you have drawn upon is the intellectual property of an institution with clear guidelines on representation, do you feel that there are any dangers of misrepresenting such research if presented through art objects and in a museum context?

CY: Yes, which is why I'm extremely careful about collaborations. There are a lot of conversations that run in the background where I discuss the project with my collaborators (be they scientists, chefs, artists, students, companies, etc...), why I'm doing it, what possibilities could follow once we exhibit the work in public, as well as an opportunity to say no to further involvement. I send out updates from time to time, and I amend all my websites regularly so everyone knows what I'm up to. I've been on fellowships and grants for a long time, so I understand the importance of being accountable and making sure everyone is on the same page; that way we can all continue on happily. In terms of the information and images I send out to the press or publish online, I always make sure everyone involved has had a chance to review them and has no issues with the material. I think life is really short and there is no sense in prolonging suffering, so if either one of us is unhappy, and all possible solutions have been exhausted, I probably will end the collaboration and just change the direction of the project.

That said, though, in my experience, if the finished project is successful – and I define "success" here as when a project gains an audience, when the message about climate change is effectively transmitted to another person, and when the collaboration was pleasant and we want to do it again – everybody wins: the artist, the collaborator, the space it was exhibited in, the audience who has had a positive experience. If the project fails at any point, such as if a blogger completely misinterprets it, it's only me that has to bear it. In this situation I try to rectify it by reaching out to the writer with more information and an invitation to get in touch. Either the writer updates his article or I'll be working on another project to further make the point. The public events are really critical to me, and I usually have attendants – I call them The Apocalypse Squad – who are trained to talk about the project and assist the audience if needed. The online presence of the projects is also important because that's where I put all the information. I'll get the occasional troll in the exhibition or on the Internet – usually climate change deniers – but I just ignore them.

Catherine Young is an artist, scientist, designer, explorer, and writer whose work primarily explores human perception and its relationships to memory, creativity, and play. Her work combines art and science to create stories, objects, and experiences that facilitate wonder and human connection. Her first solo exhibition was in a science museum. She received her degree in molecular biology and biotechnology from Manila, fine art education from Barcelona, and has an MFA in Interaction Design from the School of Visual Arts in New York where she was a Fulbright scholar. She has been on residencies and fellowships in New York, Barcelona, Seoul, Singapore, and Manila.

Atlas Remeshed

An Interview with Davide Angheleddu

Interviews with Innovators: You recently participated in the Hybrid High-lights exhibition held at the Seoul National University Museum of Art. Two of your artworks featured, each associated with an important Swiss research center: CERN and the Human Brain Project. Without going into technical details, can you tell us briefly about these projects and the data used?

Davide Angheleddu: My **Augmented Reality** (AR) piece "Atlas Remeshed" is an artistic view of the so-called "ATLAS experiment" being conducted at the Large Hadron Collider (LCH) accelerator at CERN in Geneva. The experiment aims to identify new sub-atomic particles, or other new phenomena, that can tell us more about the basic forces that have shaped our universe since the beginning of time, forces which will also determine its fate. The specific experimental event that I have artistically re-interpreted sought to confirm the existence of the Higgs Boson, a sub-atomic particle associated with the mass of all elementary particles in the universe. It is named after Peter Higgs, one of six physicists who predicted its existence in 1964. The experiment consisted of observing two protons colliding at high energy and then measuring the trajectories of the particles generated in the collision. Depending on the kind of sub-atomic particles involved, the 3D trajectories will differ, and so can be described as having a particular "signature". In the experiment in question, the so-called "Higgs Signature" left behind after the collision consists of four trajectories, those of two muons and two electrons.

An instance of this signature is depicted in Atlas Remeshed, along with the trajectories of other particles resulting from the collision. Looking more closely at the artwork, the muons and electron trajectories are represented with strong, vivid colors (two long blue tracks and two short blue tracks), whilst other trajectories are represented by light red tracks within a transparent mesh interpolated between them. The mesh is achieved through a process called "remeshing"). Although Atlas Remeshed was created using actual data from the CERN experiment, the remeshing algorithm was employed to reveal the different proximity relationships amongst the trajectories described. The process transforms the data into a new, single geometric object that changes the event visualization to better reveal the presence of the muons and electrons.

A similar remeshing process was also used for the project "Brains Out", a second AR artwork from the same exhibition that I developed together with Arthur Clay and the artist K-Soul. In this work, the remeshing process was applied to a series of complex 3D models of neurons provided by the Human Brain Project. Here, the reprocessing reduced the 3D models in complexity and made them much more suitable for use in an AR environment. In comparison to Atlas Remeshed, this work allowed for much more artistic freedom in its creation and presentation.

IWI: In what format did you obtain data from CERN and from the Human Brain Project? Were there any guidelines or restrictions involved in their usage?

DA: I was able to obtain the data I needed directly from the scientists concerned – for which I'm very thankful: data for Atlas Remeshed came from CERN, whilst data for Brains Out came from the Human Brain Project project at Lausanne Polytechnic (EPFL). For Atlas Remeshed, I asked the scientists for data represented as XYZ coordinates, which described the 3D model of the collision event containing all the tracks of the various sub-atomic particles that I needed for the creation of the artwork. Similarly, all of the data from HBL was in a standardized format consisting of complex polygonal 3D meshes describing the shape of neurons.

I received permission to use CERN data from Prof. Sergio Bertolucci, Director of Research and Computing at CERN. The use of complementary materials such as ATLAS images and explanatory material comes under the general terms of use of ATLAS images and videos. Since I am using the ATLAS materials for artistic rather than commercial purposes, I was allowed to use them as required. However, in order to have Intellectual Property Rights (IPR) recognized for my work, without negating those belonging to the ATLAS experiment itself, I had to sign an agreement with CERN stating that I would provide a citation that clearly recognized the correct IPR for both parties. It reads: "Artistic view of Higgs Boson event from ATLAS experiments data. Courtesy of CERN. Artist: Davide Angheleddu."

IWI: Professor Gabriele Guidi and his research team at the Politecnico di Milano have worked to develop computer applications that can reduce massive point cloud data to much smaller data files, a process that lies at the heart of ATLAS Remeshed and Brains Out. Could you tell us a little about this research?

DA: The research activities of the group lead by Professor Guidi cover many different areas, ranging from 3D data acquisition, reality-based 3D modeling, 3D time-evolving reconstruction, and 3D visualization in the field of **Digital Heritage** (DH). The role of the polygonal model in all of these areas is pivotal, but its construction and visualization is often difficult without a dedicated super-computer. The research that I am developing together with Professor Guidi, within the framework of my PhD research, is related to the optimization of such models. Our research goal is to make 3D modeling more widely accessible through low-cost computers, tablets, smartphones etc... To accomplish this, we are focusing on the integration of "traditional" 3D modeling methods with techniques commonly used in the gaming industry (such as representing 3D geometries on 2D maps). Although the process is conceived initially for DH, it can also be applied to any complex 3D model made out of polygons.

IWI: How have these techniques found application in the ATLAS Remeshed and Brains Out projects?

DA: Let's take Brains Out as an example. The Human Brain Project supplied me with what is called "cloud point models" (3D models) of human brain cells. These had to be adapted in order to reduce the density of the cloud point data, so allowing 3D meshes to be generated onto which textures from K-Soul could be applied. The ability to do this automatically is quite difficult, but is one of the key elements in making this project come to life. The meshing process for adapting the original data to a format compliant with the Augmented Reality platform consists of three phases:

1.   Transform a dense cloud of 3D points into a high density mesh, i.e. a surface made by small triangles (polygons) that connect the raw 3D points as "nodes" in a polygon network;

2.   Optimize the high density mesh to reduce the polygon count by at least 10 times whilst maintaining the geometrical structure of the original as much as possible;

3.   Unfold the mesh to generate a flat plane (like unfolding a box to place its 6 faces onto a single flat surface). You can imagine this process as "de-structuring" the 3D model in order to obtain a flat version of it. The flat version is then superimposed onto 2D planes that, when stitched together, form a much simplified 3D model.

The second step described is particularly important in the development of AR applications because only models of a low complexity can be used. This step can be achieved in one of two ways:

The first is to apply an algorithm that deletes nodes in the 3D model where they are very close to each other and therefore don't add much to a description of the model's shape. Imagine how you would simplify a 3D model of a teapot: You would keep a high density of nodes in the complicated spout in order to maintain its shape close to the original; the teapot body, however, is much simpler (with lower curvature for example), and so can be described with much fewer nodes.

The second approach requires a modification to the topology of the model. A model's topology is described by how nodes can be joined together to describe a flat surface (called a polygon). In the case of developing AR applications, a process of working out how these interlinking polygons can be greatly simplified to reduce the number of nodes further, whilst preserving the global shape of the model, is required. Once created, a new low density mesh (with new topology), can be projected onto a high density mesh to allow its geometrical accuracy to be improved. The result is a new 3D mesh object with a clean topology and with a small number of polygons.

This process raises the question of how much precision is lost in comparison to the original scientific data. You can imagine the polygon reduction process as equivalent to the reduction of pixels in a digital picture. If you reduce the number of pixels, you obtain a lower quality image because the image contains less detail. In the same way, if you reduce the number of points in a 3D model you obtain a lower-quality 3D object. However, with 3D meshes you have more freedom in "moving" pixels around to best recreate the form of the original. Thanks to the simultaneous consideration of geometrical and topological properties, and the re-projection of the simplified mesh on the original one, I have been able to obtain a simplified mesh with minimal deviation from the appearance of the original: polygonal reduction without loss of quality. At the moment we have obtained a reduction of 90% in term of computer memory usage without loss of geometrical information. If AR platforms start to implement these new forms of mapping, this could be a very convenient way for representing detailed 3D models in Augmented Reality applications.

IWI: How have these different 3D modeling techniques come to inform your artistic practice?

DA: I graduated with a degree in Architecture from the Politecnico di Milano. Afterwards, I worked in an architecture studio that was specializing in interior design. It was here that I developed a deep knowledge of 3D digital modeling tools. During this time, I developed a concept of digital representation which I understand as being not only a technical tool for rendering a project but also a means of project development in which progressive refinements are made from an initial seed to create the form desired. This approach serves now as the basis of my artistic production. For me, it has been very valuable in coming to an understanding that 3D modeling can be a creative tool for visually prototyping artwork. In most of my work I combine the use of 3D CAD packages and 3D modelers used for animation films. CAD offers more rigor and geometric precision, but 3D Modelers are more agile in creating complex shapes that resemble more closely those of the natural world. This approach has developed into a kind of "3D-software cocktail" that allows me to draw natural shapes while maintaining a high degree of metric control. I have to confess that my engagement in this doctoral research began with my artistic vocation and the idea of experimenting with new techniques for creating and representing 3D models of artworks. I was aware early on that any developments in these 3D modeling tools would impact my own personal artistic research, as has been the case in ATLAS Remeshed and Brains Out.

IWI: The work ATLAS Remeshed was a physical object before it became a virtual one to be viewed in an Augmented Reality browser. What is most significant about the relationship between the real and the virtual in your artwork?

DA: The physical version and the Augmented Reality (AR) version of Atlas Remeshed are different expressions of the same digital model. I find the physical model fascinating because its development is the result of a completely digital pipeline (or work process), one that results in a 3D file that can be printed out as a plastic object, which, in turn, can be transformed into a metallic object through the "lost wax casting" technique. I find this particularly interesting because the final result comes out by an amazing synthesis of divergent technologies: the first is a new one made possible by the latest digital technologies; the second is an ancient one that has not and changed in 3,000 years.

The AR model of Atlas Remeshed is a bit like magic. It allows everybody to access the artwork with simple and widely available devices such as smartphones. This type of presentation has a number of advantages over the physical model of the same work. Scalability and positioning are easily obtained: You can create a gigantic sculpture in front of a building by just changing a configuration parameter. Geometric variations and object transparency can also be set-up during installation, making the expansion of the narration with animations and other multimedia aids possible. Currently, however, AR technology is constrained by a number of technical limitations around memory, processors, and the applications native to devices for interacting with AR content. This underlies the need for new methods, such as those I have developed, for enabling the scalability of any model with a certain level of complexity.

IWI: ATLAS Remeshed was well received during the Hybrid Highlights exhibition. Many visitors commented on its beauty, while others found it more informative than beautiful. Is it an artwork or a visualization of scientific data? Where do the two merge, and where do they separate?

DA: The work consists of two components: the tracks collected during the experiment, and the organic, transparent volume obtained through a creative process. My choice to fuse and group all the red tracks with a smooth organic volume highlights the shift from the scientific to the artistic. These choices are the result of my personal aesthetic vision. So, my artwork concerns both the pure representation of an experiment (containing all the information necessary for an interpretation by scientists) and the artistic interpretation of that representation.

The work has different depths to it: It starts by attracting the attention of the observer with something aesthetically pleasing, before progressively drawing the observer into the story behind the artwork's shape – research into the origin of the universe and everything in it. The artwork moves the spectator's attention from science to the humanities, from the experiment itself to human life, so helping him build a relationship between the artwork and his curiosity about the origins of the universe.

IWI: Atlas Remeshed and Brains Out reveal an important role for collaborations formed across different disciplines. Where do you think the value lies in creating cross-disciplinary work?

DA: I think that this type of cooperation is absolutely a consequence of the times in which we are living. A century ago, when Einstein developed his theory of relativity, it was feasible to revolutionize science working alone, as he did. Today the world is very different: Leading researchers need to work in groups to conduct experiments and test theories. I think that the same trend is perceivable in other domains outside of the sciences, and art is not excluded from that trend. In addition, interdisciplinary cooperation can be mutually enriching, a synthesis of different perspectives bringing about new and unusual points of view. As such, these interactions can support the creation of new solutions more easily because it can drive the integration of different areas of knowledge and offer a larger picture of the question in hand. What is commonplace in one area of knowledge might be a novelty in another and offer just the right inspiration needed.

Davide Angheleddu studied Architecture at Politecnico di Milano in 2000. He developed a concept of digital representation during his first design placement that acts not only as a technical tool for rendering projects, but also offers an approach to project development based on progressive refinements from an initial seed to reach the form desired. This concept has been the basis of much of his artistic output since. In 2005 he began to develop a concept of digital art that is inspired by shapes from nature. Examples of his work have been shown in several international art exhibitions, including London, Milan, New York, and Santiago de Compostela.

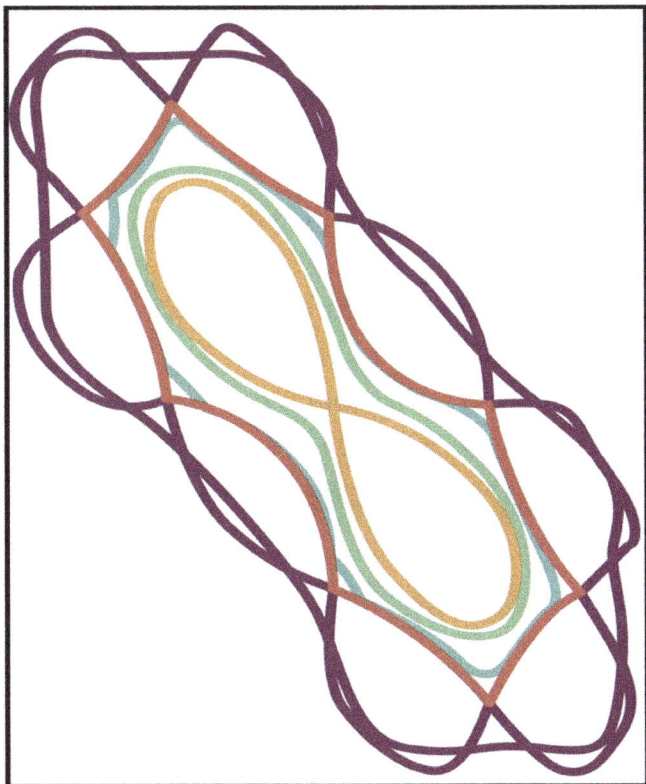

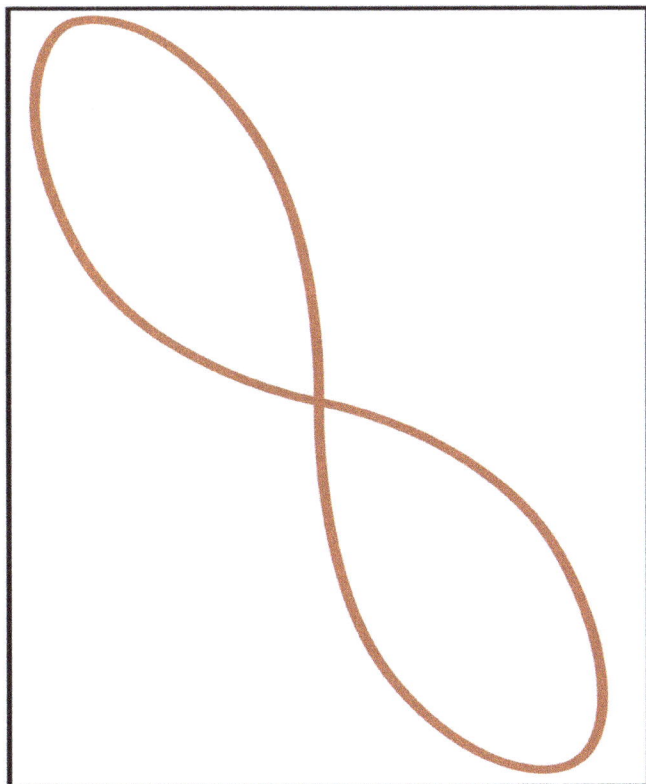

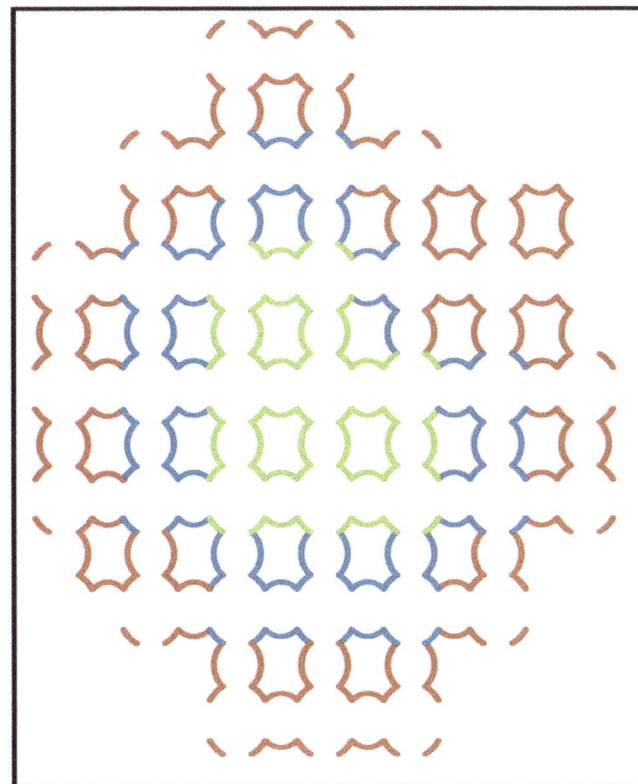

A Full Understanding of Things

An Interview with Ruedi Stoop

Interviews with Innovators: The study of aesthetics has traditionally been a topic for philosophers. How can mathematicians help us approach this subject in a different way? Further, if we say that the universal perception of aesthetics is based on predictability, complexity, and computation, are we saying that mathematics brings us closer to an understanding of truth and beauty than the arts?

Ruedi Stoop: There used to be a unison between philosophy, mathematics, and to a somewhat lesser extent natural sciences. If you need an example, take the divine section (da Vinci, Dürer) or logic (Leibniz). I see it as this: Mathematics is the last step in a 'full' understanding of things. This does not imply that the first steps need to be taken in terms of mathematics, but it does implies that the last step be mathematics. Examples abound from the fields of logic, biology, and even modern painting.

Let me state that I don't think that arts can make us understand truth or beauty on the level that I seem to reserve for 'understanding', but I do think that we can arrive at a mathematical information science formulation of the 'truth' that might be seen to underlie beauty. An understanding of things is always based on fundamental principles or notions (such as mathematically defined predictability, complexity, and computation). The basis of good science in my view is exactly measurability; a key task of mathematics (and also of theoretical physics) is to define different concepts to be measured in an objective way. I am even tempted to say that notions that are not anchored in a measurement process should be banned from natural science (despite the many enemies that such a statement will generate).

Only if we are able to elaborate the fundamental principles that rule aesthetics will one be able to properly understand why some directions of art will persist and others will not. This is a quite general remark. Among the fields I work in, phenomena that I and my students have treated in a similar manner, are power laws in networks, fundaments of hearing and pitch perception, and noise in biological information processing, to name but a few. The description of a phenomenon from fundamental principles must finally result in an expression through a measurement, because otherwise it will be useless. The extent to which the measurement supports experience with the object then tells you how good the model is and how it should be adapted if you are going to refine it further.

IWI: Complexity emerges in the interaction between humans and their environment. How do you understand the aesthetic experience as a function of that complexity?

RS: Complexity emerges not just 'at' the interface between humans and the environment. High complexity emerges whenever we have real interaction between different physical phases, properties, or things: It is the interaction itself (if this happens). In common language, therefore, complexity comprises formulations like being able to influence a variety of possible future developments, maintaining or being able to track a memory of the past... Aesthetics emerge where complexity challenges the human mind or intellect. This is where it comes from, and this is what it is designed for. In that sense, aesthetics can be seen as the reward to the human mind or intellect upon answering this challenge. Depriving the human mind of being able to have this experience means to dehumanize it. There are many psychology studies that support this observation.

IWI: How does this relate to the early work of D.D. Birkhoff who proposed a mathematical theory of aesthetics based on the relations of elements – rotation, symmetry and so on – as a measure of complexity within basic shapes?

RS: The work of Birkhoff regarding aesthetics pointed in the right direction, but suffered from a containment imposed by classical geometry that he seemingly was not able to get rid of. This is remarkable since he can be seen to have had all the ingredients to understand the matter from a very much more fundamental and unconstrained angle than he finally did. Of course, symmetries have a great impact on complexity that lie at the foundation of art, since it defines one basis from which the intellectual challenge of prediction starts. However, fully symmetric objects are generally dull. This has been long known in the practical 'aesthetics of beauty'. Aesthetics is the interplay between new rules or symmetries violated. It is this game of discovery and violation of rules that challenges our intellect (which can happen via a range of diverse pathways) and not the mere existence of a symmetry per se. In a sense the 'failure' of Birkhoff's notion of aesthetics seems to me similar to the failure of Kolmogorov's algorithmic complexity in its application – mostly unanticipated and unintended – to natural complexity.

IWI: How does your model of complexity differ from Kolmogorov's algorithmic complexity, and how does this better help us to understand the complexity that arises through our interactions with artistic work?

RS: Given an object, Kolmogorov's approach starts from its description via a single 'trajectory' (imagine a perfect drawing of the object using a pencil). You then estimate how many bits you would need to store this information, using the most efficient compression program available. This is the nature of Kolmogorov's complexity. For human use, however, such a description is generally more precise than needed; the human mind simply does not work like that. We look at things mostly through ideas and concepts, i.e. in a more abstract way. Our objects of perception focus on 'ensembles' of similar 'trajectories'. Kolmogorov complexity is inspired by how a classical computer computes (although it is generally impossible to find the most efficient compression program), but this results in an inadequate reduction of the world compared to how we see it. As an example, look at a trajectory that is generated by a random process.

In terms of Kolmogorov's complexity, this process has maximal complexity; there is no way to describe real randomness other than by the chain of the events it produces. Seen from a human – or, if you prefer, artistic – point of view, a two-dimensional random process generates just dull, non-complex, pictures. From an ensemble point of view, a very short description of such processes can, however, be given. For example: 'Throw the die until your arm falls off and then look at the result as a set'. Art, working essentially with the tools of abstraction, can be seen as the perfection of the human way of looking at things.

IWI: How would you outline the steps required to get from a mathematical model of complexity to a model of aesthetics?

RS: The key step in obtaining a model of aesthetics is to have an appropriate measure of complexity. In our work that defines the complexity of prediction – prediction is the fundamental task of an intelligent system such as the human mind – we measure, in simple terms, surprise, i.e. the failure of prediction. Surprise is what fuels the human mind; I think it would be possible to give arguments from evolution for such a statement. To approach human perception of the world and the perception of aesthetics computationally, we need to start from ensembles of trajectories (or from the 'ideas' that are expressed by these ensembles, if you prefer). From these ensembles we infer how difficult it is to predict the further evolution of the ensemble (imagine that we gradually discover the nature of a painting). The surprise that we are confronted with during this process can be measured, and this defines the object's complexity.

Fully random or constant processes yield, of course, zero complexity. The most complex processes are the intermittent ones, which nicely ties to the classical observation that biological complexity dwells at the border between order and chaos. Intermittent systems are confined for an untypical time within one 'world', so that the prediction or anticipation of another world from within the present one becomes virtually impossible. Such properties are valuable for living systems in the sense that this reflects the compromise between their ability to adapt quickly and the necessary retention of memory at the same time.

IWI: How might the mathematical approaches you have described best shape, or influence, the work being done in other fields such as neuroaesthetics?

RS: I have to confess that I had to look more deeply into how people working in these fields understand their aesthetics (remember that I mainly work as a mathematician/physicist in the field of biocomplexity). What I have seen so far did not really align with my interests. First of all, where the sensation of aesthetics is located in the brain is of as much interest to me as is a road map of London if I have to stay at home. This may help to understand the brain on a certain level, but for an understanding of what aesthetics is I would prefer a deeper understanding of the matter at hand, and to be able to formulate this in terms of mathematics. A second group of people working in the field has tried to comprise in words what the properties of aesthetics need to be. I feel that a formula of the kind I offer, one that can measure the aesthetic quality of a work based on its underlying complexity – as the state between total order and total noise, but still explorable to the human mind – is much more powerful than any verbal description. For example, the role they generally attribute to 'symmetry', when looked at it from my point of view, is entirely and verifiably false.

IWI: If aesthetics is an expression of the human mind's desire to perform computation – where the larger the challenge that can be mastered, the larger the pleasure the human mind gets – then could one say that this mathematical approach doesn't interest itself in questions of artistic style or genre? Do you see a role of art history and art criticism in making sense of how complexity arises in experience with art?

RS: Well, I am a mathematician and a physicist, an information scientist and perhaps also a bit of a neuroscientist, but certainly not a specialist in artistic styles or genres. Although I am personally not disinterested in the arts, I am certainly not qualified to discuss art history or art criticism. Of course, the questions of predictability that I claim underlie aesthetics are embedded into society as much as culture can be described as being powered by the 'Trieberfüllungsverzicht' (S. Freud). In this context, and in addition to the fundamental percept of aesthetics related to predictability of patterns, there are the avalanches of ideas that an object might trigger in your mind. What it exactly triggers depends strongly on the socio-cultural embedment of the observer's existence.

As a person, you need to be open to other people, towards things, and towards options in order to be able to experience aesthetics. Curiosity is, therefore, one of the pillars for being able to experience or fathom aesthetics; if you can't see it, or more probably don't want to see it, you won't really experience it. I would probably be willing to go as far as to say that, in my understanding and experience, I doubt that there could be such a thing as a contemporarily Stalinist or fascist aesthetics. Aesthetics is the opposite of a totalitarian view. So, while it is conceivable that, in the advent of fascism, objects of the so-called 'pure art' had this triggering property, the same objects (e.g. sculptures) appear vacuous and dull from a different socio-cultural embedding fifty years later. Although their appearance may even be, and often is, formally perfect, the expression by the pattern alone is just not inspiring enough. That is because, seen from today, they fail to make a deeper connection to the human mind, and they fail to inspire meaning and thought. One is tempted to say that being able to generate a perfect copy of nature without the ability to more deeply reflect the fundamentals of life – in terms of estimation and prediction – does not appear to most of us as aesthetic. Looking at Greek statues that express in their (dis-) proportions the supremacy of thought over body-power seems to express this to me, despite the presence of a usually perfect body.

Ruedi Stoop studied mathematics and physics at the University of Zurich (UZH) and the Swiss Federal Institute of Technology in Zurich (ETHZ), obtaining a Ph.D. from UZH. His PhD-thesis became a major part of a 'Springer highlight in solid-state physics' book (written by Peinke, Parisi, Rössler, and Stoop). After a habilitation at the University of Berne focusing on phase transitions in dynamical systems, he joined the newly founded Neuroinformatics research institute run jointly by the University and ETHZ, where he became a professor. Since then, he has published books in the field of nonlinear dynamics and on methods of computation. Recent work has included the first electronic hardware construction of a Hopf cochlea, and the first physics model of pitch perception that can explain all known nonlinear pitch phenomena.

Neuroaesthetics: Beyond Mere Beauty

An Interview with Jaeseung Jeong

Interviews with Innovators: The field of neuroaesthetics seeks to identify the neural foundations of the aesthetic experience. It is an ambitious goal for a relatively new field. Do you think an understanding of the aesthetic experience is within our grasp?

Jaeseung Jeong: This is a big question, and one I often come back to as I pursue my research. To get a better sense of what is possible in the field, let's look at an example of the type of research we're doing in my group. As you know, background information is typically provided to viewers when viewing artworks in a gallery, a museum, or an auction house. But how that information influences aesthetic judgments is still not known. People tend to like what they know, with an accompanying ease of visual information processing.

When background information provides the viewer with an understanding of the artist's intentions, this helps to distinguish meaningful figures from abstract forms; it may increase the familiarity of the viewer with the artwork presented. So, we recently examined whether increasing the viewer's knowledge of abstract paintings changes their subjective aesthetic valuation of such work. Subjects were exposed to five pieces of information (such as artwork title, artist name, critic commentaries, and a valuation figure), and were asked to repeatedly reappraise the work.

We classified this information into two categories based on our hypothesis. The first type of information provides an understanding of what the artist has tried to express through the artwork. We expected that knowing the artist's intention might increase familiarity with the artist by easing the viewer's perceptual burden of interpreting abstract forms. In contrast, the second type of information describes features of the painting that would attract the viewer's attention to those areas of the artwork that exhibit high semantic congruency with that information. We compared the viewer's initial aesthetic judgments with changes in attention after multiple reappraisals. In addition, we used an eye tracker to examine patterns of attention exhibited by subjects, and determined the neural correlates of aesthetic judgment during artwork reappraisal through using functional magnetic resonance imaging (fMRI).

We observed that knowledge-based aesthetic judgments are coded in the orbitofrontal cortex (OFC) and the rostral anterior cingulate cortex (rACC). Neural correlates of increased aesthetic judgment based on each type of information showed the involvement of discrete brain regions associated with sensory and attention processing during reappraisal. The artist's commentary decreased activity in the visual cortex and the visual associative areas in the left hemisphere, indicating that the alleviation of perceptual difficulties related to the abstract expression allows increases in aesthetic judgment processing. However, the critic's commentary also directed the viewers' attention to specific visual components in the painting, activating voluntary attention networks in the brain. Cognitive information modulates a viewer's sensory brain activity; our results tell us that this modulation depends on the characteristics of information provided, and that the newly updated aesthetic values are integrated into the activity in the prefrontal brain regions reported in association with value-based decision making.

I am not sure whether an understanding of the neural basis of the aesthetic experience is within our grasp or not. However, I believe that these kinds of results and interpretations are not possible through traditional aesthetic approaches; that is the unique selling point of **neuroaesthetics**. It will definitely expand our understanding of the aesthetic experience.

IWI: The importance of cultural and social factors in shaping the aesthetic experience has begun to receive more attention in the field in recent years. Does this herald a new phase for cross-disciplinary work in neu-roaesthetics?

JJ: This is a very important issue, one that we are currently working on. Again, let's look at a particular example. Let us consider the situation that, thanks to new media, a consumer can access reviews such as that provided by a Yelp reviewer. These reviews offer expertise that guide consumer choices. A number of previous studies have shown the strong impact of other people's opinions on our own judgments, attitudes, and value preferences. The influential power of experts have traditionally been employed to guide consumer choice. New media practices now support increased access to the opinions of others, and marketing experts have investigated the impact of peer influence on consumer behavior. Although experts and peers have been suggested as two major sources of influential opinions, the mechanisms of this influence and how they incur changes in one's judgments have not been directly compared at the neural level. Do our brains give the same weight to the opinions of peers as to those of experts?

When individuals become aware of the judgments of others, they may revise their own judgment to adopt the same point of view. There may be a number of reasons for this: Individuals often believe that the decisions of a group of people may be more accurate than the decisions of a single individual, and may thus follow the opinions of others; the opinion of another person may provide knowledge or indirect experience that one had not previously considered; the potential rejection from a group, and the possibility of humiliation by others, may make an individual adjust their initial judgment in order to conform with the group. However, comparing the value judgments of experts with those of peers has not been investigated with regard to the brain's social processing areas. Additionally, it is unknown which of those brain regions involved in changes of judgment based on the known opinions of others (e.g. through access to a reviewer's opinion) may be distinguished from those brain regions involved in changes in judgment aimed at mitigating conflict with a majority opinion (e.g. through access to group rating scores). I think there are so many significant questions that remain to be answered in this area of research.

We should note that recent neuroimaging studies have demonstrated that social influences activate the brain areas associated with the processing of social conflict, the ability to adopt another's perspective, and self-referential processing. Moreover, the magnitude of the activations could predict subsequent changes in judgment; the dorsal anterior cingulate cortex (dACC), for example, has been shown to become activated when individuals realized that others had a differing opinion. Previous studies have also demonstrated that conformity to a different perspective may be modulated by the opinions of experts, a small peer group, as well as by a majority view. The brain areas where activity was found to correlate with subsequent behavioral change, as a result of persuasion by others, included the medial prefrontal cortex (mPFC) and the right temporal parietal junction (TPJ). This is a good demonstration that a neuroscientific approach can provide an insight into the impact of cultural and social factors in shaping the aesthetic experience. New cross-disciplinary approaches to the study of aesthetics will, I think, begin to emerge.

IWI: In Neuroaesthetics, a tension arises between the search for common processes that we all share and the aesthetic moments that we each experience individually. Where do the strengths and weaknesses lie for a reductionist account of the aesthetic experience?

JJ: Universal or shared features of aesthetic experience based on neural mechanisms discovered in the field of neuroaesthetics might be very fruitful for our understanding of the aesthetic experience of artworks. Individual variability in the aesthetic experience (arising during very special encounters with artworks in a gallery or museum for example) which deviates from universal features will also become an important topic in this field. I believe that neuroaesthetics can provide us with insight into such questions as "why do we create artworks?" and "why do we enjoy artworks?" Furthermore, neuroscientists are beginning to explore a greater range of artworks that would allow us to ask more varied questions about why people are drawn to art, and to explore the richer types of experience that people have with art. For example, I am working on exploring the neural processes that arise in response to contemporary art forms, including the work of Jasper Johns or Damien Hirst. The number of scientists working in this field with an interest in contemporary art will increase in the near future; they will expand these studies to address both a greater range of art forms and the types of encounter we have with them. As contemporary artists are not interested in merely perusing beauty, this research area promises to become much more interesting and varied.

Jaeseung Jeong is an associate professor at the Department of Bio and Brain Engineering, Korea Advanced Institute of Science and Technology (KAIST). He received a Ph.D. from the Department of Physics at KAIST in 1999, and subsequently became a postdoctoral Fellow in the Department of Psychiatry at Yale University School of Medicine, and an Assistant Professor in the Dept. of Psychiatry at Columbia University College of Physicians and Surgeons in New York. His interests include human decision-making, brain dynamics, and brain-robot interfaces. He has published several books on science that are national bestsellers in South Korea, China, Taiwan, and Hong Kong.

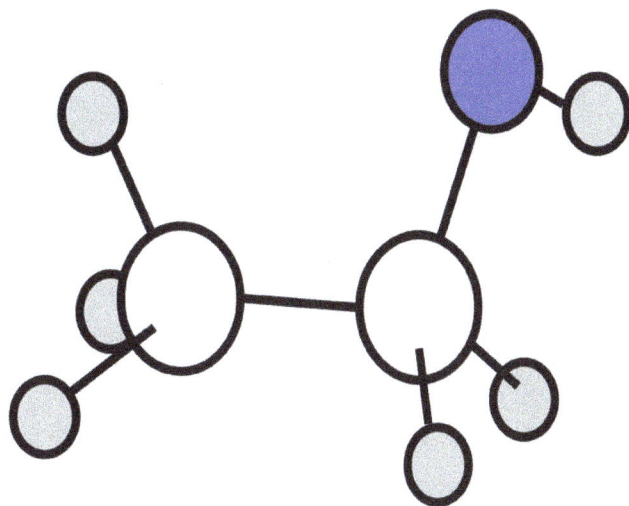

No Science without Humor

An interview with Dr. Anatophil

Interviews with Innovators: You are a professor of anatomy at Ajou University School of Medicine, but also a full-time cartoonist exploring new ways of teaching anatomy to students and a wider general public. By way of an introduction to your research, could you tell us about the Visible Korean Human project and how new digital technologies are finding use in your field?

Min Suk Chung: Since 2000, the Ajou University School of Medicine has been working in collaboration with the Korea Institute of Science and Technology Information to produce serially sectioned images of the whole human body. In this process, a cadaver is cut at 0.2 mm intervals using a cryomacrotome, each time revealing a surface that can be digitally photographed to produce an anatomic image. These digital images reveal not only high resolution detail about the structure of the human body, but can also be used to create virtual reconstructions of hundreds of different individual structures within the body. The result is a collection of stereoscopic models that can be selected in any combination and freely rotated in real-time for in-depth viewing. Furthermore, the realism of these "solid" models can compensate for many of the limitations of the flat anatomic images produced. The results of this work have been made available in a variety of different formats including PDF files and software applications for viewing digital materials.

Anatomy is a subject learned by students majoring in medicine, dentistry, and healthcare science. Through these 3-dimensional models they are now able to experience a cadaver dissection virtually, and so gain a rich encounter with the human body that they can repeat again and again. For those interested, more information on the Visible Korean Human project can be found at anatomy.co.kr (where materials can be accessed for free).

IWI: How did you first got into the study of anatomy, and what led you to become the cartoonist, Dr. Anatophil?

Min Suk Chung: I graduated from medical school to become an anatomist, selecting the scientific path rather than the clinical path. I was promoted to full professor at 45. My work at Ajou University is simple: I teach and I research anatomy. After I received tenure I wanted to turn my attention to actively revealing the inner workings of anatomy and science in a more entertaining way. For that reason, I chose the medium of the cartoon. I have become well acquainted with numerous amusing stories about my job over a period of almost 30 years. After finishing a scientific experiment I'll write an article. Now, after hearing a funny story, I'll draw a comic strip.

The very first cartoon I created about anatomy was in the year 2000. This stemmed from an initial motivation to make medical students aware of certain anatomy-related situations that become amusing only once one has acquired the requisite knowledge. They are then better able to recognize how enriching knowledge of anatomy can be! Dr. Anatophil is cast as the main character in many of my comics. Using myself as the model – we have similar attributes after all – helped to set a lighter tone and make the comic strips more enjoyable for my students. Although I have attempted to make my English cartoons known internationally, I don't think they are known outside of my home country. Dr. Anatophil is, however, well known in South Korea.

IWI: You create comic strips to explain anatomy in humorous ways. What do you see as the link between humor and science?

MC: In my opinion, scientific presentations, whether verbal or written, need a sense of humor. This is because a good presentation must possess all those aspects that will make it interesting to other scientists; a scientific article that has a sense of humor is more easily accepted by a journal, and is more likely read when published! When I'm dining or drinking with colleagues, I'm really not very serious at all; I often make jokes about my scientific activities. What makes me laugh the most is science that reminds me of cartoons, and cartoons that remind me of science. The former refers to science that is stimulating and draws you in like a good cartoon, the latter to cartoons that have the same logic as good science. I think my work shows that science and cartoons have many things in common.

Humor is also a great way to get students more involved in their studies. In class, a student may not really care about a lecture they've just missed, but when you hear someone laughing at the punch line to a joke you've not heard yourself, then you want to find out what the joke was: "Why are you laughing?". Humor interests students; it can draw their attention. I think humor is significant in science because it can make this difficult field of study so much more approachable. If you take a look at a few examples of my comics, it is easy to see how my cartoons contribute to science education.

IWI: Is it an imaginative idea or scientific knowledge that comes first in the making of a cartoon?

MC: I had an active sense of imagination as a child, often posing questions such as: "What would it be like to be a superhero?" Now, as an adult, I still use this powerful kind of imagination in the creation of my cartoons. In terms of where my initial ideas come from, I don't really invent the story lines at all; in most cases I collect ideas from what I hear around me. Commonly, I get ideas during private conversations with my colleagues and put them down on memo paper; they inevitably end up in a comic strip.

IWI: Once you've come up with a new cartoon, do you first share it with your colleagues in the scientific and academic community for feedback, or do you just trust your gut feeling and publish?

MC: I always share my comics, and not only with colleagues but also with others through the Internet. I put comics on Facebook and get feedback from my friends and acquaintances. Usually I can expect them to give feedback, which is mostly positive. In general, the South Korean scientific community is very accepting of my cartoons. For instance, I am asked every year to give presentations about my comics at conferences. I also think that, here in South Korea, we respect people who have a strong individual character. For example, the Korean Association of Anatomists recently accepted one of my reports to be their feature article, a work titled "Evaluation of Anatomy Comic Strips for Further Production and Applications" (Anat Cell Biol 46: 210-216, 2013).

IWI: *Cartoons are powerful visual messages that convey immediate visceral meaning in ways conventional texts often cannot. Can your cartoons help both students and a wider public to learn about anatomy?*

MC: I see the primary purpose of these comic strips as aiding medical students in memorizing the complexities of anatomy. Students have confirmed to me that the comic strips serve this purpose, but I have also measured the educational benefits of cartoons directly. My comics were evaluated by 93 Korean medical students (60 males and 33 females), all of whom knew about human anatomy. Before beginning a series of lectures, printouts of 650 South Korean comic strips were distributed among the students with the recommendation that they read them. After the anatomy class had ended, the relationship between the anatomy scores and their exposure to the comic strips was measured. This showed that the anatomy grades were better from the medical students who had read the comic strips. My anatomy comics have also helped medical trainees to discuss anatomy with a general public that lack a medical background, as well as with family and friends. It really is possible for a wider audience to understand my cartoon episodes without a prior understanding of the field. Through these cartoons, they don't only learn about anatomy, but also get exposure to real disciplines involved in real anatomy research.

IWI: For those who are interested in your comics, where can they go to see more?

MC: To learn more about my cartoons, you can visit my website at anatomy.co.kr. Registration is not required, and it's all for free. New episodes are added on a regular basis, and their number is steadily increasing. So, if you are a typical PhD student and need to laugh about the burdens of being one, or you are interested in comics in general, my suggestion is to visit the site when you have some time and see what is new in the world of comics and science!

Min Suk Chung is a professor in the Department of Anatomy, Ajou University School of Medicine, South Korea. He received his B.S., M.S., and Ph.D. degrees from Yonsei University, South Korea. He became interested in virtual dissection after obtaining his Ph.D., going on to perform preliminary experiments for the "Visible Korean" project. He launched Visible Korean in 2000 to improve the quantity and quality of the initial "Visible Human" project. He wishes, in the near future, to form a virtual image library that can support convenient and fast access to anatomic images for downloading. This will include all data sets from the United States, South Korea, and China. He is also a cartoonist, publishing cartoons about anatomy.

Art, Science, and The Brain

An Interview with K-soul

Interviews with Innovators: You created an Augmented Reality installation based on research data from the Human Brain Project for the Hybrid Highlights exhibition. How did the idea of creating an AR visualization of brain cells come about?

K-Soul: The idea of a project began when I was approached by Arthur Clay, an artist who has been promoting and curating new works using Augmented Reality for about three years. He started the Virtuale Switzerland in 2012 with the intention of showing virtual artworks, but, finding very few artists in Switzerland active in this area, turned his attention to Swiss artists whose work might instead be transposed into an Augmented Reality environment. I was one of the artists asked to participate, and he assisted me technically in creating or adapting my work into this virtual space. We devised a project called "Brains Out", and proposed to add textures derived from my light paintings to the 3D models developed from brain cell data acquired from the Human Brain Project (HBL) project at Lausanne Polytechnic (EPFL). It would be an inter-disciplinary project a la "Art on the Brain". We all liked the idea as the project grew out of our mutual interest in trying to create a kind of virtual walk into the brain.

IWI: What were the different roles you each played in the project, and what kind of collaborations were needed to make the project work?

K-Soul: Realizing the project required an ongoing cooperation between the Digital Art Weeks Festival and the Human Brain Project, with Arthur Clay meeting several times with the EPFL team responsible for content creation and marketing. When Arthur and I got together at Jardin Cosmique, we worked out the possible scenarios for creating an art piece at the Museum in Seoul, and I, like the HBL, made content suggestions for the new piece. HBL choose the actual Neurons that would be used, and I choose and developed the textures that would work best with those forms. However, it was the LG Company in the end that added the final piece to the puzzle by offering us a set of their new curved screens for use in the project. The screens were setup to show my light paintings from which the brain cell textures were derived, and to show a newly created film by the HBL that showed never-before-seen data visualizations from their research on the human brain.

IWI: Your project "Jardin Cosmique" is a space in which you live and work. As a whole, the design of the space is based on the anatomy of the human body. Could you tell us more about Jardin Cosmique and what motivated you to create this space?

K-Soul: The common ground in the Brains Out project between the research of the Human Brain Project and my own work is the study of human anatomy. The best example of this can be found in the structure of the interior of Jardin Cosmique, a unique event space whose design is based on human anatomy. The approach here is one of convergence and the development of a total artwork, a Jardin Cosmique incorporating all forms of art: poetry and architecture, painting, sculpture, installation ... this artwork expresses the fundamental structure of the human and its universal nature.

Any original artistic approach entails an inner journey in search of inspiration, one that engages ideas and creative forces. I call Jardin Cosmique the inner hearth of these creative forces, the garden of thought which spreads its light by our imagination. The awakening of this holokinetic garden inspires originality, creativity, and poetry in all human activity; it raises this act to the level of art. Integrated within us, this garden is the foundation of our humanity, making us powerful creators.

IWI: What role does scientific knowledge play in the creation of your work?

K-Soul: My artistic activity in general is driven by fundamental research into human nature and both the origin and activity of creative forces. I experience and refine this research by developing what can be termed "holokinetic" artworks which integrate poetry, new technologies and scientific knowledge. An essential step in the development of this process was the invention of the holokinetic painting, living light paintings that reveal the bright light and dynamic activity of thought. Working in this area required me to learn and master a dynamic pictorial language with the capacity to express itself through metamorphosis of colors and shapes. The laws and principles that constitute the basis of this universal language are deeply embedded not only in external nature but also in the constitution of the human body.

Scientific knowledge is essential to mastering this language. The study of anatomy, for example, allows the artist to learn about the most perfect work of art on earth – the human body – and to understand the organic processes underlying its creation, birth and life, its diseases, as well as aging and death. Thereby, artistic sensitivity is instilled through knowledge and upraised. On the basis of this knowledge, and other factors, I develop my holokinetic paintings to express the poetic life of a being of light. This profile oscillates between organic processes in the making and existing geometric structures, between inflammation and sclerosis, between combustion and mineralization. The pictorial organism can dissolve and integrate back into its own living structure by inflammatory combustion: Colors heat up, rhythms accelerate, and lines develop in curvature.

IWI: How was this approach important for the creation of Jardin Cosmique?

K-Soul: This approach finds architectural expression in the development of Jardin Cosmique, this poetic space modeled after the human being. This space is tripartite: At one end of the structure is the cranial area, where mineralizing forces predominate and are expressed by lines, angular shapes and structures of crystalline nature; at the opposite end of the space, combustive metabolic processes are expressed by rounded and twisted volumes; at the center, in-between, is the thoracic area, which connects and balances these two poles. The location of the various areas of activity within the space correspond to the location of the organic processes in a human being: The central fireplace acts as the heart space and regulates the thermal aspect; the position of the digestive system in the structure defines the location of the kitchen and bar; the library is a space dedicated to thought and reflection and is located in the cranial area; the office area, used to manage and control the businesses and the external exchanges connected to Jardin Cosmique activities, is in the metabolic pole. The humans occupying the inside of the space are seen as the blood of the space and considered as red blood cells; they represent the vectors of life and activity.

IWI: The work of Ruben Nunez has been pivotal in the development of your holokinetic artworks. Can you tell us more about this artist and his influence on your work?

K-Soul: Ruben Nunez was a Venezuelan artist born in the 1930s. His work is considered to be the beginning of the "Kinetic" movement. Rising to prominence in the 1950s, this movement was then taken up conceptually at a later point by artists such as Soto, Le Parc, Pol Bury, and others. In 1974, Nunez laid down the foundations for a concept of "Holokinetism", a term Nunez created by combing the Greek words "Holos" (all, whole, unity) and "Kinema" (movement). Holokinetism fused both art and science to emphasize the role of new technologies in exploring the use of light in creating artwork. Nunez worked in collaboration with the Institut d'optique of Paris and the Holographic Institute of New York; he is considered to be a pioneer in holographic art. According to Nunez, Holokinetism links fundamental research about the universe, matter, energy, and light, while also making use of the pictorial language of poetical expression developed essentially by expressionists of the time. Simply stated: Holokinetism expresses the creative power of the light.

I was working on developing a similar approach to making art when living in Paris at the beginning of 2000, one which can be experienced in the living light paintings that I am now creating. Perhaps it wasn't by accident that I had come to be living at the home of one of Nunez's best friends. This friend, looking at what I had created, said: "That's incredible. During the seventies Ruben spoke to me all the time about the use of screens as canvas, in order to integrate light and movement into a painting and to create a poetical cosmos of light." He then told me that I was doing exactly what Nunez had spoken to him about, and, although he did not know if he was still living, would try to call Nunez in Caracas. He called and found that Nunez was very much alive. Hearing about my work, Nunez answered him: "Don't move, I have to come and meet this Helvetian guy personally and see his living light paintings." That's how Rubin Nunez and I met, and when he saw my artworks he exclaimed: "That's it, that's the future and I have it in front of my eyes. Incredible!"

Following our encounter, we lived and worked together for a couple of months in Switzerland. Here we developed the concept of Holokinetism further, creating a film about these ideas. After his time in Switzerland, Nunez went back to Venezuela. We were still able to continue our work on Holokinetism, and enjoyed a rich correspondence about the merging of art and science, and how new technologies could be used in accordance with the evolution of art. Ruben Nunez died in Caracas in January of 2012. Although he was given a National Arts Award in 1959 for his glass designs, he never received recognition for his work on Holokinetism. He was a visionary, but has remained quite unknown.

Honoring him, I have adopted the term Holokinetism to describe my artworks. I feel that it fits them perfectly; I believe that the dynamic poetic and pictorial transcript of the metamorphosis of colors and shapes in nature and the universe which form the basis of the concept of Holokinetism continues intact in my light painting works and sculptures.

K-soul is an artist who develops work at the border between art and science. He allies modern technologies and traditional techniques to realize holokinetic artworks, poetic light gardens which he titles "Jardin Cosmique". His work is a reflection of, and is focused on, the human being, the origin and essence of creative forces, and the holokinetic aspect of the universe. K-soul has shown his work in various exhibitions as well as in galleries dedicated to his work, as in Barcelona (Jardin Cosmique Barcelona from 2010 to 2012), in the Principality of Andorra (Jardin Cosmique Andorra), in the Swiss town of Montreux (Jardin Cosmique Montreux), or at Laboratoire Jardin Cosmique in Fenalet, Switzerland. He has received the distinguished Award of Excellence from the Bauhaus Universität in Germany and the Premio Award of the Florence Biennale.

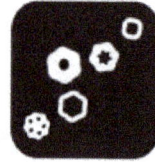

ON ART

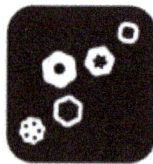

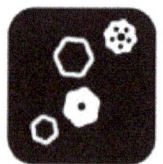

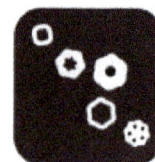

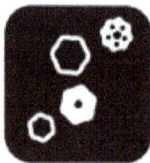

Artists have always responded to the world around them, re-appropriated the past through new means in order to create new ends, and used the lens of society's current opportunities and challenges to project possible futures. In our hyper-connected, technologically-enabled, curiosity-driven world, the emergence of new points of connection with science, technology, politics, society, and the economy is inevitable, but also high unpredictable. If Art is Life, should it aim to bring all onboard and then disappear entirely? Can an orchestra of young musicians unite a peninsula driven apart by political process? As artists move closer to science and technology, what chance does serendipity and disorder stand against hypothesis, model and method? Can we still find salvation through art?

산

울 자 자 옥 찰 자
서 지 지 바 초
의 바 한 경 산 여
부 부

의
여자

한
옥

바지

부츠

그림자

TV

지붕

BABELLISSIMO

Dubbing and Dubbing It All

An Interview with Simone Carena

Interviews with Innovators: You are an Italian architect living and working in Seoul, South Korea. Much of your practice focuses on how ideas and content can be translated in new ways to break down established boundaries. What was your very first project as an architect?

Simone Carena: I believe architecture is order. The first time I tried to organize my things or thoughts and give them space in relation to each other was when I started the "architect-thing". To have a talent for orientation and the ability to think three-dimensionally are certainly advantageous. My first design project was a rational layout for my mother's new kitchen. It was, above all, a scenario-based, production-chain-oriented, and ergonomic-centered design where the fridge, removable vegetable table, wash sink, cutting area table, trash cabinets, cooking stove, tools, electricity, gas, preparation table, and dining cart were all designed with the correct heights, materials, lights, and sizes. The first large project I undertook was for a cemetery, a project won through a competition by my first architecture office. I was just out of grad school.

IWI: Was there a difference between the idea of becoming an architect and the realities of practicing as one?

SC: The difference is dependent on whose perspective you are taking. If it is the client's idea of what an architect is, you can end up producing quite ugly spaces and still be praised as an artistic, eccentric archistar. If it is your idea, and it is detached from reality, you can end up just hitting a wall or somehow surviving through ego alone. However, I believe that being an architect has to do with keeping relevant, because architecture was born with things like tombs, which are supposed to last longer than one generation. Fad and Trendy architecture, as well as interior design

67

and decoration, are the noise in efforts at making a more relevant foundation. To think long-term is vital to our profession: If superficial architecture is like the weather, then real architecture is like the climate. I have been in South Korea since 2001, and many of my friends who have tried to work here gave up after about two years. This is because it is a brutal market, but, in defense of effort, the longer the battle lasts the sweeter the victory.

IWI: MOTOElastico is a small company working in a very busy sector with many large players. How does this affect your trajectory and what you value in your architectural practice?

SC: MOTOElastico's role in Korea – and from within Korea – is really a form of orbiting-trajectory control. We negotiate influence, gravitational pull, and the danger of collisions. We take space very seriously so we rely on our sense of humor to face the dark matter of the construction business; we find massive clusters of stars attractive, but also dangerous.

Most large commercial architecture offices produce dull spaces; MOTOElastico dreams of stupendous ones and recreate samples in our microcosm. We are aware of our minuteness in the scale of outer space, but we value conscience with science. But, don't get me wrong, we like beauty. Unfortunately, many social projects and community-driven designs are just plain ugly! They're good ideas, but they look very bad. There is a social value in beauty and it is not linked with price, but with quality.

IWI: When a project is developing well, how do you keep it moving forward? Further, do you think satisfaction with the end result can be planned in advance?

SC: This year I taught a rather strange class at Hongik University about satisfaction in design. From the discussions with students, I learned that we don't think about or plan well enough for satisfaction. Rather, we follow our instincts and "blend in with the tribe" more than we exercise our ability to act on a dream and plan for the realization of that dream. We join clubs, join religions, and follow brand fads that provide us with ideas of happiness. But happiness is a dynamic process and follows a curve that must have both low points as well as high points if it is to stay relevant in our lives. Bracing for a low bounce can be rewarding if you hit the bottom in the right posture and can enjoy the rise back up. Also, the roller-coaster moment of zero gravity right before the drop is very exciting – if you trust the coaster's engineering. I believe that a solid and consistent life structure offers great roller-coaster rides, but that must also includes periods of quiet … waiting for your turn, chatting with and meeting other riders.

IWI: Looking at your sketches, a lot of progress in your work can be seen from one to the next. How important do you feel the act of sketching is to the development of a project?

SC: Sketching is becoming more like making memos from a flow of thoughts, which is not much different from doodling while talking on the phone. The ability to connect different things, mix unexpected ingredients, and to find links where there aren't any to be seen is the key to the process that we use. It works the same way that synapses work inside the brain: "neurons that fire together wire together". We just simulate this process at the conscious level through testing new associations between the things we can see, hear, smell, touch, and so on. New juxtapositions between these ingredients result in new ideas. At MOTOElastico, we talk about trajectories, crash landings, gravity assists, and escape velocities when discussing a topic, but one can use other analogies as well. Comedy is a good example. Riding the paradox and delivering a punch line is pure creativity: something new bursting out of shared premises. Good comedy takes you to new and interesting places through a strong structural framework.

IWI: Architecture becomes boring when it lacks originality, but also offers no comfort when it is too novel and lacks familiarity. How do you keep a sense of familiarity in your work while still remaining original?

SC: If I say a building is "boring" it is because I was expecting to be entertained by it; if I say a building is "familiar" it is because I can appreciate the sense of home that a comfortable space can deliver. Glamorous architecture becomes boring quickly and it is a contradiction to the very nature of architecture, because architecture is something that has always had to last over generations. Today, it is just a newly made face of a trending brand. The most tiresome buildings of my generation are those using screen-projection. With these "you can project whatever you want onto their screen-like facade". However, it is duller than it is flexible. The effect is just like that of a recurrent fashion show with naked models where "fashion is said to be surpassed and the body is now the key". Projected facades work well only at night, and, with the high energy usage required and the endless looping of material, they get both expensive and boring very quickly. In contrast, a sober pyramid, a complex temple in Angkor, a Roman villa, or even a 1930's Le Corbusier building don't become instantly redundant and boring. They are alive both inside as well as outside; even an ancient tomb is full of life! None of these buildings need either LED lighting or the inane use of light projections.

As already mentioned, architecture should be a long-term experiment, one that must be able to manage a lot of irrational, unscientific, or unpredictable inputs. So, the true goal is not originality per se, but rather finding an effective new combination of ingredients that can form the start of new experiments, experiments that will, in turn, spawn originality.

IWI: FUBU in Myeongdong was designed by MOTOElastico. It was a truly eye catching design which included a life-size airplane. Walking down Myeongdong Street it certainly got noticed. What were the goals of that project, and how did the inclusion of airplane elements come about?

SC: Fubu is a great story, and it's filled with irony. Originally, FUBU was a NYC Brooklyn-born, black rapper street wear brand that was bought out by Samsung. Once the brand became Korean, it completely lost all of its street-wise, badass New York City soul and became a cosmetic, metrosexual, cutie cutie, K-Pop boys band costume. So FUBU constantly lost ground in the market place. We were hired to design the last flagship store for the final resuscitation attempt of the brand. The client's interest was to try and revitalize the musical roots of the original branding. We suggested an airplane-cargo look, where everyone in the store would be part of the shipping crew responsible for transporting the sound system, the instruments, the clothing of the stars on tour etc... Basically, the customers would find themselves under the plane and inside the cargo belly, i.e. placed in the role of grassing through the stars' personal wardrobes. In the beginning, we had an actual landing gear system from a Korean Air Boeing positioned at the entrance to the shop! It was massive and rather beautiful, but – despite it all – it had to be removed at the last minute because some prudent Vice President of the company believed on the basis of feng-shui that it would bring the business a lot of bad luck. I guess it's also never a good idea to remove landing gear at the last minute, because after that incident the brand ultimately crashed. Today, in the place where the FUBU was located, there is yet another cosmetic store with a predictably boring façade. (What we actually need is cosmetics for facades, or maybe even cosmetic surgery!).

IWI: *What advice can you offer students concerning where they should look for inspiration and how they should realize their ideas?*

SC: I think they should look in places that are largely unrelated to one another and construct a kind of conceptual bridge between them. They should look into how things work, check out the etymology of things to discover the beautiful codes of meaning that lie beneath, search the net and use it as an image source, and above all let things lead them astray. Put your favorite things together and remix it all, ask somebody to taste it, think about it, express what you like about the things you hold close, discuss with friends and debate with enemies, and listen to good music (of course).

IWI: Tell us bout your thinking behind DUB? Where did it originate, and how do you use this concept to shape your architecture?

SC: Originally, dubbing was a technique used to manipulate existing music recordings with the aim of producing new instrumental remixes. This was often achieved through significantly reshaping the recorded material, for example, by removing the vocals and emphasizing the drum and bass parts. For us, we apply these techniques to architecture; to DUB is to translate and tell original content in a new way. We use dubbing as an approach to re-design spaces, especially to remix traditional Korean spaces. The process is not accurate, but it is passionate and can impact society through triggering new questions around present and future culture. Some examples of our work here include the Hanok House Dub, the Happolice Installation, and Seoul Citizen Hall. The Hanok Dub project from 2008 is the manifesto of "Dub Architecture", a MOTOElastico trademark. Here, we used modern materials and our Dub manipulation style to remix a traditional Korean house (a Hanok) into a contemporary home, blurring the boundaries between interior and exterior, old and new. A second example is the Happolice Installation, from 2009, developed in collaboration with Oksang Lim. Here, we transformed former military buildings at Kimusa (Seoul) into a new, politically charged performance venue. As part of the redevelopment, the walls of the building were removed, the interiors painted pink, and a 112 pink police shields mounted onto the façade. These projects include work for both private and public spaces, reveal a range of aesthetic interventions, and capture different types of engagement with social and political discourse in the public sphere.

IWI: One of your colleagues, Moon Hoon, has said that a good architect is a crazy architect. What makes a good architect in your opinion?

SC: Craziness is a complex condition (from the dictionary I like definitions 2 and 3: "crazy informal 2 extremely enthusiastic: I'm crazy about Cindy | a football-crazy bunch of boys. 3 (of an angle) appearing absurdly out of place or in an unlikely position: the monument leaned at a crazy angle."); craziness is an impassioned and alternative take on life. Moon Hoon is not crazy. He is very smart and represents salvation for the Korean scene because he is international and domestic at the same time. He is designing and building for people, not brands.

Concerning what makes a good architect, I was recently asked to imagine a manifesto for the future of cities (Seoul 2024), and after some thought I imagined the evolution of translating technology and how that will make all barriers obsolete. Not only languages but also brands and financial lingo will be revealed and opened. We will be able to rebuild the city of Babel with towers that can finally reach God, peacefully. Isn't that both a plausible future and a great one to pursue?

I never really understood why the God we know from the Bible was so angry with his children, who were all speaking one language and working together to reach him (or her). To aspire and be good is the usual teaching, but some form of ferocious jealousy kicks in if one ever gets too close to achieving it. It's like saying: "Let the children come to me", but not so close because I am wearing silk.

In the end, I like to think that the new Babel will be a celebration of understanding between cultures and creatures, not a race to surpass God but be a common effort to reach Good.

Simone Carena is founder of MOTOElastico, an architecture office he runs with Marco Bruno. It is known as the best Italian architecture office in South Korea. He is Assistant Professor of Digital Media and Space Design at IDAS, Hongik University, Seoul. He received a Master of Architecture from Southern California Institute of Architecture (USA) and a Master of Architecture with Honors from Politecnico di Torino (Italy). Further studies have included Architecture and Design courses at Oxford Polytechnic University (UK), Harvard Graduate School of Design (USA), and The University of Technology of Kingston (Jamaica). The architecture project "Hanok Dub" was published in the Architectural Review, C3, and the New York Times. The project was awarded First Prize in the 2010 Asia Interior Design Institute Association (AIDIA) Residential Building Unit Competition.

Augmented Increments

An Interview with Felix Heisel

Interviews with Innovators: "Augmented Increments" is a project that seeks to challenge the way we think about social housing and community building. Could you tell us more about the project and what it involves?

Felix Heisel: "Augmented Increments" is an experimental social housing project that aims to provide partially constructed shelters in low- and no-income zones which inhabitants finish over time according to their needs and financial means. The name Augmented Increments is a combination of two important aspects of the project: Augmented Reality (AR) and Incremental Construction. The second aims to activate homeowners in social housing projects by incorporating their skills and capital into the construction process over time. Our tool incorporates the use of Augmented Reality, aiming to visualize this potential for stakeholders and decision makers. A physical model of the project describes the basic structure to be completed, whilst overlaid digital images (viewed through a mobile device supporting an AR environment) shows the impact of different materials and construction methods, suggesting how such a structure could be populated and used by its inhabitants over time.

The project was developed by myself and researchers Dirk E. Hebel and Stefan Müller-Arisona at the Future Cities Laboratory (FCL) in Singapore / ETH Zürich in collaboration with Sheer Industries Group Singapore. Augmented Increments is part of the interdisciplinary research project Addis2050, which includes researchers from the Ethiopian institute of Architecture, Building Construction and City Development (EiABC) in addition to those from the FCL.

IWI: At the heart of the incremental housing concept is the idea that only the basic infrastructure of a house is constructed before it is handed over to the owners. Have you put these ideas into practice, and are similar models to be found elsewhere in the world?

FH: The Ethiopian institute of Architecture, Building Construction and City Development, together with the Bauhaus University in Weimar, have recently finished a prototype of their "Sustainable Incremental Construction Unit" (SICU) within the setting of an informal settlement in Addis Ababa, the largest city in Ethiopia. It is a two-story building which provides a finished room on the upper level and an unfinished, open ground floor. Future owners or tenants of the building can then decide how to finish and use the space, be it as a family home or a property that can generate some form of income. The building is still uninhabited at the moment, but time will show how it is going to be adapted.

Incremental housing models have also been implemented elsewhere. Since 2004, the architecture office Elemental S.A. in Chile has been constructing several projects where only one-half of each row house is initially built. Afterwards, the owners or tenants of the house can decide to extend the house underneath the already existing roof in accordance with their own capacity, identity or aesthetic desires. Such a scheme minimizes the investment volume through small, flexible housing typologies, and helps to elevate home owners into the role of urban house builders through following clear sets of rules concerning health standards, material specifications, accessibility, and fire regulations.

IWI: Is there evidence that this model can lead to sustainable housing development?

FH: The examples in Chile not only indicate that such models work, but that they even increase the economic and social value of such housing programs. Compared to large governmental projects, self-build housing concepts usually have a much smaller grain size, which complements the life style in low- and no-income areas. The concept of self-construction also increases the identification of the owner with his home – an important precondition for guaranteeing care and maintenance. Lastly, that these spaces can be multifunctional is key to their success; it allows small-scale businesses to operate within these social housing developments, and so provide income for most of the families that live there.

This element of the flexible adaptation of one's living and working space over time is very important for the development of successful housing concepts. Informal housing settlements in developing countries have clearly shown that the inhabitants possess the skills and means for self-construction. Incremental housing concepts aim to channel already existing construction activities into a legal and infrastructural framework by providing basic structure and planning. By meeting in the middle between formal housing provision and informal housing construction, the goal to achieve an economic and social win-win situation for both parties involved can be achieved.

IWI: What building materials are available in the low- and no-income areas you target, and how does this impact your conception of an incremental housing project in action?

FH: Informal settlements have two main material sources: locally available building materials (such as natural stones, loam, timber or grasses) and recycled materials (which includes metal sheeting, doors, windows, and building elements formally used in other structures). Recycling provides the foremost material resource for self-build constructions, and so has been taken into consideration in the design of incremental housing projects. The different processes attached to the use of recycled materials also usually result in jobs and incomes for many people in these settlements, and thus play an important role in their everyday lives.

IWI: With a focus on affordable and personalized construction, is the incremental housing concept also applicable to other housing markets?

FH: UN Habitat defines no-income zones as areas operating on an income of less than $1 per day. Such settlements appear mostly in developing territories and in urban agglomerations. However, incremental housing concepts are not restricted to developing areas; a recent design project created for the International Building Exhibition in Hamburg, for example, experimented with similar ideas but in the context of a wealthy German city. In the end, economic factors such as construction cost and rent prices will decide on the success of such design concepts. In the end, economic factors such as construction cost and rent prices will decide on the success of such design concepts. In developing territories, the need, however, to provide shelter for an ever-growing urban population will play an important role in the decision processes that will shape these housing markets; self-construction might be the only chance we have to provide sufficient homes for everyone.

Felix Heisel is currently working as a researcher in architecture and construction at both the ETH Zürich in Switzerland and the Future Cities Laboratory in Singapore. Preceding this position, he was the coordinator for the 3rd year architecture program at the Ethiopian Institute of Architecture, Building Construction and City Development (EiABC) in Addis Ababa, Ethiopia. Felix recently published the book Building from Waste: Recovered Materials in Architecture and Construction (2014, Birkhäuser), and has contributed articles to magazines and books such as S.L.U.M.Lab: Made in Africa (2014, UTT), The Economy of Sustainable Construction (2013, Ruby Press) and Building Ethiopia: Sustainability and Innovation in Architecture and Design (2012, EiABC). Felix has won several awards over the last few years, including the Ministry of Education (MOE) Innovation Grant (2014), the SMART Innovation Grant Singapore (2013) and the Bauhaus.SOLAR Award (2012). His interest in Ethiopia's urban growth resulted in the making of the movie series _Spaces, a collection of six documentaries on the appropriation of space in Addis Ababa started together with Bisrat Kifle in 2011.

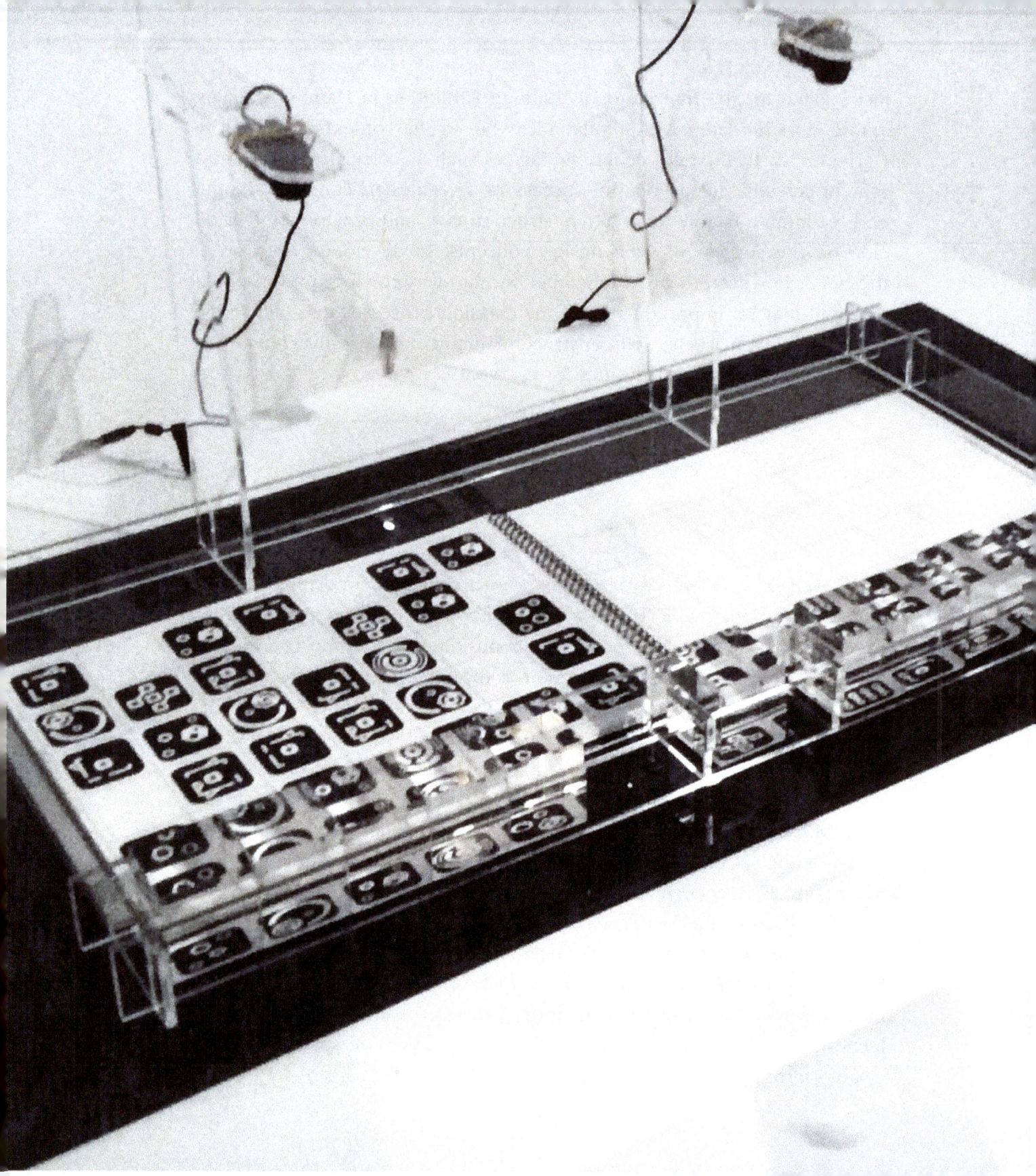

The Museum and You

An Interview with Boa Rhee

Interviews with Innovators: To begin the interview, could you tell us a little about your background and your role at Sogang University?

Boa Rhee: I studied museum management and marketing at New York University and at Florida State University. I eventually returned to South Korea, where, for the last 14 years or so, I have been participating in policy-making decisions in the Ministry of Culture, Sports and Tourism and acting as a consultant for a number of museums. In my current role at the Department of Art and Technology at Sogang University, I teach art history and aesthetics, media planning, and the management and marketing of cultural institutions and culture consumption. Museum studies is not integral to those classes, but the skills and technologies required for museum management and marketing are strongly emphasized. The curriculum includes a wide variety of topics. These include writing exhibition grants and proposals, designing and developing exhibitions using new media, conducting visitor research, and the use of mobile technologies. I also place great emphasis in my courses on how approaches and understanding from the arts and humanities can be used to promote creativity as a key element in developing new ideas about museum management and visitor experience.

IWI: How important are these ideas around museum management and visitor research in South Korea Today?

BR: The number of museums, including but not limited to art museums, in South Korea is approximately 600. However, South Korean museums lack an awareness of the importance of management and the need for visitor research. With regards to museum management, every one of these museums faces the same problems around funding as their counterparts abroad. Some museums in the private sector face serious difficulties in developing exhibitions and educational programs as they simply lack museum personnel and cannot afford to hire new staff. To solve some of these issues, the Ministry of Culture, Sports and Tourism is currently considering the development of a museum accreditation program that could be implemented at a national level. This aims to promote the importance of museum management in creating successful and sustainable museums for future generations.

IWI: In technology use, we see the opportunity for museums to have an engagement with their visitors that extends beyond the museum visit itself. With reference to your research, can you tell us about the concept of the "seamless visit"?

BR: The model of the "seamless visit" takes into consideration not only the actual visit to the museum but also those periods pre- and post-visit. Sherry Shi coined the term "seamless visit", but other terms have also emerged (Silvia Filippini-Fanton & Jonathan P. Bowen call it "the virtuous circle"). Within the context of this model, visitor research focuses on understanding the full experience of the museum visitor, from their prior interest and understanding about an exhibition, through their interactions with the exhibits, staff and other visitors during their visit, to the post-visit discussions and activities that may drive further visits in turn. All this information is very powerful in helping museums best respond to visitors' interests and expectations; it will help them encourage visitors to return again and again.

The model also emphasizes the use of enhanced interaction between the visitor and the museum's real, as well its virtual, interfaces. As such, the model opens up the use of modern communication technologies such as smartphones and tablets. The ubiquitous and pervasive nature of today's communications networks, when combined with mobile devices that are easily portable and intuitive in their use, allows museums to integrate technology into all three periods defined in the model, so allowing bridges to be built between them. In a museum setting, such technologies becomes particularly powerful; they can be used to shape the location-based and context-sensitive situations that are most significant in how a museum's collections are experienced.

IWI: The use of technology to view and interact with artworks in an exhibition is becoming more common as a way of enhancing visitor experience. What are the opportunities and challenges that museums face in adopting technology in this way?

BR: In my understanding, interaction can be defined in terms of the relationship between the exhibit and the visitor. Interactive technologies open up new ways in which visitors can meaningful experience objects on display. When properly implemented, interactive works can lead to a better visitor experience and increased opportunities for learning. Nowadays, museums offer visitors many ways of interacting with exhibits. This ranges from hands-on engagement with artworks through to touchscreen computer displays. These encourage visitors to delve more deeply into the context of objects on view. With this in mind, we can see how interactive technologies can be useful tools. They are not only useful for making art but also for enhancing visitor experience. Mobile technologies are also capable of compensating for certain spatial and temporal restrictions that arise in the design of exhibitions, as well as helping to "remove" the physical boundaries that separate the inside from the outside of the museum.

However, as technology becomes increasingly embedded in museums (operating in both physical and virtual contexts), it is critical that a museum's management understands how visitors behave towards interactive objects, and grasp what the underlying causes of those behaviors might be. This is essential because museum professionals often misunderstand the needs and expectations of visitors. There are, for example, many reasons why museum visitors may prefer not to engage with interactive exhibitions; understanding an artwork can be a challenge in itself, and participating in exhibitions that require engagement through interactive technologies can make that challenge less appealing or even inaccessible. What this underlies is the importance of implementing effective visitor research and developing methods for collecting and responding to visitor feedback.

IWI: What measures should be taken to help museum visitors get the most out of interactive exhibitions?

BR: The biggest challenge for a museum is to encourage visitors to engage with those artworks that use interactive technologies in the first place. Central to this is training staff to teach visitors how to engage with those artworks in a way that reduces the resistance they may feel towards them. In this way, the role of the museum attendant is beginning to change. In a traditional museum setting, museum attendants largely help visitors through offering spatial directions and answering general enquiries. In interactive exhibitions, however, the roles of these museum personnel changes to focus on providing visitors with the information, tools and support they need to experience the exhibit in question as fully as possible.

In an interactive exhibit, visitors can often feel that they are not permitted – or that it is impolite – to use their mobile phones in a museum context. To get visitors to engage with artworks through their phones, museum staff needs to communicate the value of this form of engagement just as they would if they were encouraging visitors to explore the artworks without technological help. Equally, without clear explanations, visitors will have little, if any, idea on how to interact with exhibited artworks. These issues become more acute when dealing with visitors who either rarely go to museums (and so might not be used to this type of technology application) or come across exhibitions by chance (and so might be less prepared for what they are about to encounter). A final, and equally important, role that the contemporary museum attendant must be able to carry out is to solve basic technical problems with interactive exhibits when they arise. This includes, for example, being able to handle connectivity and software application issues.

So, we are seeing that the role of museum attendant is changing from a "passive" role to a more active role, one in which a degree of technical understanding and good communication skills – commensurable with the interactive exhibition in question – are needed if museum visitors are to get the most out of their museum experience.

IWI: With reference to exhibitions such as Hybrid Highlights, do you think the added dimension that Augmented Reality works can provide will encourage visitor engagement and increase both the diversity and size of museum audiences?

BR: I have a few reservations about the use of Augmented Reality in exhibits. I spent about an hour observing the behavior of visitors exploring the AR exhibits at Hybrid Highlights; whilst I noticed that young visitors were able to actively participate with AR objects in a short period of time and without any hesitation, other types of visitor seemed more reluctant to engage – despite the presence of museum staff on site for guidance. This is most likely because they didn't fully understand the concept of Augmented Reality and had never experienced its use in an exhibition before. Some were also faced, as is common, with technical problems. As discussed earlier, I believe that the careful planning and management of technology-use in museums can solve such issues if considered solutions are placed in the hands of museum staff. If these conditions are met, there is no reason why the use of new technologies shouldn't be able to attract a broader visitor base to interactive museum exhibitions.

Boa Rhee was born in Seoul, South Korea, and has a doctoral degree in Museum Management from Florida State University. She is currently an associate professor at Sogang University, South Korea. Her main interest is the Smart museum platform and the creation of the seamless museum experience through the use of mobile technology. Her work focuses on the exhibition technologies, management and marketing strategies, visitor research, and research into wearable technologies and mobile devices. Boa Rhee's book "Museum Management and Marketing" has just been published, and she is currently working on a second entitled "Mobile Interpretation: museum informatics and mobile technology".

Accu

Keeping It Healthy and Curious

An Interview with The Curious Minded

Interviews with Innovators: The Curious Minded is an international arts group with a guiding principle of keeping healthy and curious, whilst remaining anonymous. How would you describe your group in your own words?

The Curious Minded: Basically, it's an arts group, but not really a group of artists. It is more like a network or pool of willing souls who are interested in doing something under the protective guise of an alias. Members come and go, and any idea we have is passed from one to the next like a baton in a relay race. We kind of think of each other as tools for transformation, but we don't realize the works ourselves; we prefer to offer galleries, museums, and festivals instructions for the realization of our work. If we don't make things then we don't have to show up at exhibition openings.

IWI: How did it all begin?

CM: It all began in Berlin on Käthe Kollwitz Platz. There is a playground there, and at that time it was not so the "olive-eating-wine-drinking-designer-baby-carriage" scene it is now. It was still a little rough around the edges, and there was still a glimpse of East Berlin left. Anyway, the first piece was a trophy work. It was a snowman which was made and then stored in diverse refrigerators, only to be scooped out time and again to make Vodka frothies.

IWI: How did the group expand from a couple with a snowman into an international network of art collaborators?

CM: Sharing refrigerators can be a bonding experience, and if you add a little Vodka to that you get pretty willing collaborators. Actually, it was low key and people just asked the curators who we were and if they could get involved. So they just simply asked, and if we liked them – and they were keen to follow "the rules" – they were brought in and became members.

IWI: As a group, you are trying to stay off the "beaten path". Has this been difficult to sustain?

CM: It must be possible, but it seems that when things aren't mainstream, or haven't gone pop, they aren't given any ramp time. It is a sign of the times, unfortunately. We think a lot about this, and, because CM's motto is "keep curious and healthy", we decided first and foremost to have some fun and make our main point of creating work to keep curious. I guess people liked it and became curious themselves. So we got shows because they were smiling.

IWI: What is the most important thing you want to achieve through your work?

CM: We like to think that our currency is memories, and what makes it important is that we created them for the people involved. As mentioned, we don't attend the openings, but, if we do, we come unannounced. What is important to us then is the tickle of creating something that is completely conceptual, because in a way it has all the beauty and also the convenience of dreaming.

IWI: You have exhibited in Canada, Japan, Singapore, and South Korea. What was the most important or most memorable show?

CM: To answer your question for the reader in a more practical way, the work in Canada is something people would remember. The exhibition took place at Open Space in downtown Victoria. There was a need to create a kind of freebie for people who bought festival passes. So we provided instructions for a project called "Nine Phases to the Moon" which explored a lot of odds-and-ends around Chinese space travel. The highlight was the DYI Chopstick Trainer Kit, which, when assembled, would blink each time you took a bite using the chopsticks. Each kit had its own color, and the idea behind it was to answer the question: "What color was the moon"? The lights shone onto a paper placemat which had a picture of part of the moon, so people could see the moon change color whilst they ate.

IWI: *And what about your work for South Korea?*

CM: That was kind of a number-and-a-half. We had to work with a lot a tech people, and they were spread out all over the place. It was also a "Boot Strap Project", and so a few students were selected from Sogang University who would then realize the work. It's called "Diamonds ForNever", and the instructions were to create Augmented Reality Diamonds that could act as a viewing filter for the artworks on display. They did a great job, and apparently it was really well received because it lets you see the exhibit through rose colored glasses. Who would not like that?

IWI: *The artwork for South Korea was the most complex work the group has done to date. The works for Japan and Canada were "paper glue and scissor" works. What's next for the group?*

CM: Actually, we never really know, but I think we will just start folding stuff and forget about the glue and scissors for a while. But, on a more serious note, we kind of like to create stuff people can do something with... like play, eat, drink, or make things they can even take home. It doesn't really matter what, but we were happy about the museum in Unsin Japan which was interested in picking up the Volcano Game we designed. So, maybe some more games, maybe one on how to find a job.

IWI: *So life is art and everyone is an artist then?*

CM: Yes, and again yes. We all have to survive in the money world, but life and art are really more about staying curious and healthy. And I think that can happen if you stay fresh and keep on doing creative stuff. Here's to good health and curiosity!

The Curious Minded's guiding principle as an international arts group is simply to remain healthy and curious. Since their founding in 2010, Curious Minded has expanded with a "no-names-needed" approach. Members currently reside in Berlin, Montreal, Warsaw, Zurich, and Seoul. The group has produced a diverse number of art objects in the form of witty games, viewing cards, bedtime stories, and a variety of easy music works. Several works are now available for purchase at museum stores. Because their work lies outside of the periphery of the art market and is created to be "played", the group aims to avoid more traditional paths in the art world, focusing instead on developing new approaches to exhibiting work. Curious Minded has exhibited in museums and galleries in North America, Asia, and Europe.

Curating Infinity

An Interview with Houngcheol Choi

Interviews with Innovators: The importance of technology for a new generation of artists is central to your work as a curator. Can you tell us what is meant by your term "Techno-Romantic" when it is used in the curatorial sense?

Houngcheol Choi: I originally majored in sculpture during my university days, and I worked both as an artist and as a curator until about 2003. The first time I curated a large, site-specific media art exhibition was for "Digital Art Network" in 2001, and the success of this exhibition led me to curating the newly created biennale, "media_city seoul 2002". So, I have worked as a curator for around 17 years. In my experience, the key external factor that has dramatically changed modern art in South Korea is the development of technology. Since the late 1990s, high-speed communication, mobile technology, and personal computers have spread rapidly; Augmented Reality has become common with the proliferation of the Internet. The introduction and proliferation of digital technology has not only changed the way we live, but also the way we're thinking. The artistic position that South Korea has developed, and accepted, is allied closer to American Modernism than European Modernism, the former being more theoretical and pro forma. Under such influences, Romanticism – the counterpart of mainstream Intellectualism – has been a concept that the majority of artists have ignored.

The term "Techno-Romanticism" applies to contemporary artists, specifically the new generation of artists who are playing around with digital technologies. In the term I see an opening to address the inequalities that arise through technology usage, empowering artists and audiences to think and act differently around technology. Compared to ideas of pre- and post-European Romanticism, the present artistic situation in South

Korea offers many routes into a discussion about Techno-Romanticism. One starting point I have proposed is the idea of "Bad Romanticism", which focuses on the differences in attitude towards technology between established, traditional artists and those of a new generation who have grown up with personal computers. The latter are skilled and passionate about arts and technology whilst at the same time being able to easily absorb new phenomena, and so are able to challenge ideas from traditional practices.

IWI: In what way do the exhibitions you have created since 2001 – including media_city seoul 2002, Cool Bits (2006), Bad Romanticism (2011), A Night on Galactic Railroad (2013), Super Nature (2014) – come under the concept of techno-romantic?

HC: Much of my work draws on this concept in some way or another. The "New Vision-Digital Art Network" exhibition, for example, drew parallels with McLuhan's idea of media as sensory extension. The venue was the recently-opened Seoul Techno Mart, a twelve story electronics market housing diverse electronics shops where people would shop for the newest computers, cameras, audio devices, and mobiles, or simply to watch movies. At the center of the building was an atrium that extended down to ground level. Here, sixty artists – mostly South Koreans – installed around a hundred artworks in shops, elevators, show windows, as well as throughout the whole building. For me, the venue felt like an army storage room made with a contemporary public in mind. I organized the exhibition by simply aligning the contents of the shops with the five senses as expressed through the artworks. For instance, I connected "the eyes" to videos and TVs, the ears to audio systems, and "the mouth" to communication devices such as phones. Later, this way of creating connections became much more complex and systematic as my thoughts about the senses extended to include arms, legs, torso, brain, skin, and neurons. Similar to the original concept, I connected these parts of the body with different artworks.

At the media_city seoul 2002 exhibition was a work entitled "Luna's Flow", referencing Namjun Paik's "Moon is the Oldest TV" and Sukbosangjul's epic "Moon Shines Thousand Rivers" (the very first epic written in Hangeul). From the perspective of Techno-Romanticism, the work suggested to viewers that contemporary media should be received in a more emotional and culturally-minded way.

Perhaps a final example will suffice? "A Night on Galactic Railroad" was an exhibition inspired by the posthumous novel of the same name published in 1930 by Kenji Miyazawa. The exhibition was developed for Goheung, an island in the southeast of Korea where the Naro Space Center is located. Like the novel, the exhibition posed important questions about whether communication between natural life and technology is possible. The place of the exhibition and the actual experience of seeing it brought about changes in visitors' understanding of the topic at hand.

IWI: It is often the responsibility of a curator to design a "sense of flow" through an exhibition. How do you see technology and interactive artworks changing both the face of exhibition design and the role of the curator?

HC: I think that contemporary art that follows the development and growth of technology will become not only more detailed in nature, but also provide an increasingly multi-sensory experience over art that does not. My interest, however, is not only to enrich an individual's experience of an exhibition; I also want to create a shared experience. Whenever I curate an exhibition, the use of technology allows me to integrate spontaneous elements in order to provide the audience with an unpredictable and highly original encounter. For example, the experimental exhibition I curated called Super Nature explored how technology can be used to expose a world unseen.

I am a believer in new media exhibitions, and I think curators must understand technology themselves – as much as artists and technicians understand it – to do their jobs. A curator's job is largely as a researcher and organizer. The work we do gradually steps us into new realms of experience through interaction and cooperation with artists and technicians. In the future, curators and artists will eventually exchange each other's roles, and the role of audiences will become more prominent than even artists might imagine. This is because the public value of an artwork is not directly created by the artist; it is, rather, created and given dimension through public interaction with the work. I have to cherish the site-specific nature of my projects in person in order to be able to enjoy the creative experience with an audience.

IWI: Beyond technology, many of the artists you work with are creating what is known as "site-specific" artworks, those that respond to the unique characteristics of the space in which they are exhibited in some way. Can you tell us about how you work with artists when curating such works?

HC: In my case, the spatial characteristics of the venue are very important factors when curating. I have been interested in installation work since I was at university, and I was involved in making experimental art projects for about two years after my graduation. So, I became used to exhibiting outside of galleries and museums, experiences that first took shape in the creation of the Digital Art Network exhibition of 2001, where it was possible to design a site-specific installation from the very start of the project with the artists involved.

A good example of how my site-specific works develop is the "Dreamy Hive" project from 2003, an exhibition in which I participated as an artist. At that time, I got to know a person who wanted to renovate a two-story house as a commercial building. So, I came up with the idea to turn the house into a temporary exhibition space, inviting a diverse range of artists including painters, installation artists, and new media artists to participate. The artists and I created all of the interiors, carrying out work such as covering up windows and knocking down walls to create a truly unique space. Typically, I ask the artists not to think about fitting their works into a predetermined spatial layout, but instead to consider their own personal understanding of a space, and to reflect that into the works they were planning to show.

As such, in my experience, the only way to help an artist respond to a space is to open a discussion with them. When inviting artists to create work in this way, I always provide important information about the space and give my own thoughts about it as a venue. I do this in order to explore possible points of connection between their work and the space it will reside in. For example, when curating the exhibition Super Nature, I explained my thinking behind the title, the selection of artists I had made, and the historical background of the venue. Through the ensuing discussions, the artists and I were able to develop the concept of exhibition.

IWI: Site-specific works can occupy small spaces or cover entire cities. In the case of works installed in the public realm – in which the audience consists solely of passers-by – how does a curator communicate an exhibition to a public who has not consciously chosen to visit it?

HC: There are many approaches possible here. In a metropolis like Seoul, you can discover neon signs and LED displays all around the city; so, recently, media artists have started to install their own interactive works at bus stops or on subway platform to attract the attention of pedestrians more easily. Another example: When I curated the Media Festival in Gwaygju in 2012, I choose 5.18 Minju Gwangjang (5.18 Democracy Square) as the venue. The square is an historical and symbolic site, one where people were killed during the pro-democracy movement. Many of the old buildings in the area were soon to be re-designed as the Hub City of Asian Culture. So, to attract people to visit the Square – which was at that time already under construction, so an unlikely exhibition venue – I had to be very careful in selecting and placing artworks that audiences would not only be able to identify with, but also understand in the context of such a sensitive location. Different strategies also had to be adopted to attract the public to engage with the artworks in this public place. For example, viral marketing through social networking sites and blog articles become a powerful means to advertise the exhibition.

IWI: Given your broad interests in different types of artistic practice, in technology, and in how they create cultural and social impact when brought together, what motivates you in deciding how to curate an exhibition?

HC: My professional goal is not a fixed one, nor has it ever been so. In the exhibition Super Nature, for example, I was inspired by the politically unique history of the venue, namely the National Museum of Modern and Contemporary Art in Seoul. When curating this exhibition, I strived not to isolate new media from traditional art genres, but, rather, to integrate multiple disciplines whose separation is often down to political rather than artistic agendas. At the moment, new media exhibitions that seem to be experimental in nature are in great abundance, but they are insubstantial in their content and much too dependent on digital technology alone. I emphasized this to the artists in Super Nature because, at that time, I found myself struggling to counteract this trend.

In 2002, Jean Baudrillard submitted an essay to the media_city_seoul 2002 conference; in it he criticized the violence caused by the dictatorship of the eye. I admit that Baudrillard has also influenced my own criticism of vision. Still, as a curator trained in traditional art and art history, it is not easy for me to simply deny the catharsis that visual beauty provides through painting, sculpture, and video images; nor can I reject the curation of more traditional contemporary art exhibitions.

My approach is simple: I observe the developments in science and technology that may have some value for contemporary art, and I find myself thinking about how art may change through interaction with those developments. I research artists and other creative people, analyze how their works can produce positive impact, and explore how to either integrate those works into my projects or get those artists to participate in my exhibitions. I also think about how to attain the high goals that I have set my-self as a curator, so providing the most enriching and fitting environment for audiences to experience artwork. This is what motivates me.

Houngcheol Choi is currently a curator at the National Museum of Modern and Contemporary Art, Korea (MMCA) and Project Director of Banjul-Schale. He was awarded a B.F.A. (1996) and M.F.A.(1998) from the Department of Sculpture at Seoul National University, later studying Art Theory at the Graduate School of Kookmin University from 2011 to 2013, where he obtained his Ph.D. He has actively worked on a diverse range of exhibitions and projects since 2001, with contemporary art and media being his principal focus. His exhibitions have included "Supernature" at the MMCA (Seoul) in 2014, "A Night on the Galactic Railroad" at Nampo Museum of Art (Goheung) in 2013, and "Bad Romanticism" at the ARKO Art Center in 2011. He also worked on the 5th Seoul International Media Art Biennale (media_city seoul 2008) held at the Seoul Museum of Art in 2008.

We Are All the Dust of Stars

An Interview with Choi Jeong-hwa

Interviews with Innovators: You call yourself "AAA", which stands for "Always Almost Artist". Can you tell us what each of these words mean to you and how they tie into your artistic path?

Choi Jeong-hwa: The concept "art like art" is apart from everything, but "life like art" is a string that links everything. The artist who makes art like art is merely an expert, but the one who makes "art like life" is a human at the centre of society.

IWI: What role do you feel you play, or have played, as an artist in South Korean society today?

CJH: The position of today's artists in South Korea is somewhere between art and everyday life: They transform art into a source of life.

IWI: You have stated that "artificiality is our second nature". Is this a sign of "healthy schizophrenia" or a way to face the challenges of our contemporary nature?

CJH: Artificiality = 2nd nature = healthy schizophrenia = and a way to face the challenges of our contemporary nature.

IWI: How does tradition and the contemporary fuse in your artworks, and what do you think has been lost or gained in the process of creating work of art over decades?

CJH: I would like to cite a German saying: "Wir alle sind Sternenstaub". It means: "We are all the dust of stars".

IWI: Simone Carena is an architect living and working in South Korea who practices "dubbing", or the mixing of related and unrelated elements into a work. Is dubbing a term you would ascribe to your own artistic fusion of tradition and contemporaneity?

CJH: Tradition is something that is handed down but still in use. The objects exhibited in museums are not 'tradition' but just heritage. Complexity, fusion, absurdity, and the diversity of modern life is the present tradition. Similarly, chaos is not disorder itself, but instead shows us "infinite order". Sometimes art has to express the mix and coexistence of different rules – the joy to explore and the piety of reflection that comes from nature, science, everyday life, markets, alleys, and dumps. "Dubbing" is not only art but the link between humans, objects, and events.

IWI: On the same note, do you think that something is either lost or gained in the process of creating works of art that mix diverse cultures together?

CJH: The process of creating a work of art is almost the same as the process of a healthy child making use of his or her mom.

IWI: Recently, you exhibited at Seoul Station, creating large towers out of colorful plastic strainers. Seoul Square has a large number of homeless persons. Did this influence how you created this piece of public artwork?

CJH: Public art is only valuable when it completes what lacks, not when it ornaments accomplishments. I was surprised when I heard the story that homeless people – by the way we all implicitly agree they are the owners of Seoul Square – built a tower with plastic baskets to meditate and to heal. I thought it very Buddhist. Isn't it the perfect example of public art?

Choi Jeong-hwa was born in Seoul, South Korea. He works with a wide range of media, including video screens, modeled plastic animals, real and fake foods, lights, and wires. Inspired by chaotic and teeming local open-air markets, his work observes the attempts made by museums to preserve art in a world that is constantly changing and decaying. Growing up in South Korea in the sixties, a time of rapid economic growth and the start of mass production of disposable consumer goods, certainly left its stamp on Choi's work. His recent exhibition – entitled Truth – both embraced and ridiculed the combination of consumerism and pop culture in South Korea. He puts cheap, mundane goods on a pedestal usually reserved for art in an effort to restore the relationship between art and everyday life. This is because, as he puts it, everyday life is the stage and battlefield of today's art.

The Harmony Amongst Us

An Interview with Hyung Joon Won

Interviews with Innovators: You are a violinist and founder of the Linden-baum Music Festival, an organization that hopes to bring together young musicians from North and South Korea in the name of peace and reconciliation. When did you discover your calling as a musician with a role to play in society?

Hyung Joon Won: I was 7 years old when I went to my first violin concert. I remember it being such a curious event, and was really startled by the way the violin produced its different sounds and melodies. In later life, I think I was very lucky to rediscover the role a musician can play in society, and so I begin practicing violin with this role – and its responsibilities – in mind. Playing the violin is not only about producing sound and providing pleasure through music; it can also be about delivering a message with deep sincerity to those who are listening. As a musician, I am able to persuade people solely through the message that I hold in my heart, without the need for rational argument, logic, or even words. For me, learning about how music is composed was like discovering how the laws of gravity work. Music is a universal language, and through its melodies and harmonies we can help people to communicate with each other and recognize each other's humanity. Think how difficult it is to explain a concept such as "peace"; it is much easier to attempt it through music, and more effective as well. When I experience people communicating musically in an ensemble, and they are really listening to each other and trying to follow the other's interpretation, the result is extremely harmonious at many levels.

IWI: As part of the Hybrid Highlights exhibition you performed a series of new works for solo violin on the subject of the Petit Dance Macabre. These works were in fact transcribed, at your request, from a techno-musical artwork that involved the use of Augmented Reality. Can you tell us a little about this project?

HJW: The artwork that was on display in the exhibition was created by the Swiss arts group "The Curious Minded", and was based on a 15th century cycle of images. Taken from the "Heidelberger Totentanz", these images depict the underlying universality of death: no matter of your station in life, the Dance of Death unites us all.

In each of the images, death is visiting households without regard for social or economic status. In each, death is also shown carrying a different musical instrument. In the artistic interpretation by The Curious Minded, the original series of engravings were taken and modified to combine with contemporary scenes from Seoul Square – a place where the homeless gather in number.

Below each of the pictures was a **QR code** that could be scanned using a mobile phone. Scanning in the code opened up an audio file which began to play, accompanying the images with sound. Twelve Swiss composers were asked to compose "Petit Danse Macabre" corresponding musically to the figures in each of the pictures, pieces that included the instrument depicted in the original engraving.

IWI: How did the music from the art installation – heard through scanning QR codes with a mobile device – find its way into the concert hall?

HJW: I learned that the compositions I was hearing were generated by a kind of composing game, one similar to Mozart's dice game. Through the game, anybody could make an arrangement of the Petit Danse Macabre by re-mixing the order of the fragments provided. Each of the fragments was created for a specific instrument, but when they are combined into an arrangement a "Klangfarbenmelodie" is created (literally "tone color melody"). So, to create the compositions for violin solo, the composers had to re-write the melody specifically for the instrument (rather than having the melody played sequentially between different instruments). The exhibition curator, Arthur Clay, worked together with the composers to make this possible. The result was a collection of Petit Danse Macabre for solo violin, created by composers from three continents.

IWI: Outside of your performance work, you are also the executive director of the Lindenbaum Music Festival. What does this organization do, and what do you hope to achieve through it?

HJW: I founded the Lindenbaum Orchestra in 2009 with the simple aim of bringing young Korean musicians from both South and North Korea together through music. After the success of the Lindenbaum Festival Orchestras of 2009 and 2010, for which Maestro Charles Dutoit conducted a group of young, talented South Koreans, we were really hoping that the next concert would be a performance that also included young musicians from North Korea. We believed that through a series of concerts performed by a single orchestra made up of musicians from both Koreas, our efforts would set an example for reconciliation and nurture a more favorable environment on the Korean Peninsula, one that would promote more communication between the North and the South in a non-political way,.

IWI: You refer to this joint orchestra project as the "One Orchestra" project. What progress have you been able to make since 2009?

HJW: Over the past four years, we have been in talks with governments from both the Democratic People's Republic of Korea and the Republic of Korea. Both enthusiastically welcome the idea of a joint concert, and we hope with more support from them we will be able to push the project closer to fruition. In 2013, we were able to hold an unprecedented concert with a musical ensemble that took place in the truce village of Panmunjom, which is only a few meters from the Demarcation Line, in collaboration with the Neutral Nations Supervisory Commission that are based there. But overcoming the last few meters and reaching beyond the border is a very complex and delicate matter. Because there is hardly any dialogue, or even any open channels of communication, between the two sides, we decided to put more efforts into reaching out to the international community to help mediate and facilitate the dialogue between the two Koreas.

IWI: Where does the Lindenbaum organization find support to carry out its mission?

HJW: The Ministry of Unification in South Korea fully endorses our project and has granted Lindenbaum the authorization to contact North Korea. The arrangement that we have with the Ministry of Unification is that if we receive an official invitation from the Democratic People's Republic of Korea, travel permission – which is normally prohibited for all South Korean nationals – will be granted. Although we are making all efforts to make the One Orchestra project happen, it is still very difficult to receive an official invitation from the proper authorities in Pyongyang, especially when it is so difficult to identify exactly who the correct contact might be!

IWI: What next step would you most likely need to take in order to remove the barriers standing in the way of achieving your goal?

HJW: What I truly believe would help, would be to give a performance of violin music at the United Nations where Korean Ambassadors from the North and South are sitting in the audience. Such a concert would convey a non-political message of unification to the young musicians gathering for the One Orchestra. The performance would not bring peace, but it might bring hope by conveying a sense of peace through music to those listening. While this is just a small step forward, all of us in the project sincerely believe that it would be a valuable starting point or seed from which the orchestra could grow. Even the sincerity of these efforts made might just inspire people to spread our interests and hopes across the Korean Peninsula.

IWI: *In the performance of this One Orchestra, what works would feature in the program, and how would they reflect a need for harmony between the North and South?*

HJW: It would absolutely have to be Beethoven's Symphony No.9 and the Korean traditional work Arirang. The symphony is in four movements, and its famous choral finale has come to be recognized as a musical representation of Universal Brotherhood. Musicians from both South and North Koreas singing the Ode to Joy would really convey this message to all. Arirang is a traditional piece of music from Korea. Both South and North Korea submitted the song to UNESCO's Intangible Cultural Heritage list, where it has now been inscribed. So it is the perfect work for both sides to play together.

IWI: Several of the Petit Dance Macabre that you performed were composed by Swiss composers. Would Switzerland be an appropriate place to host the One Orchestra project?

HJW: We have received support from both the Swiss and German governments, and I believe it would be much easier and less politically sensitive to hold a concert in a country such as Switzerland. So, I think that Switzerland would be the correct place for the concert, because it is a permanently neutral power, and I think North and South Korea would gladly send their young musicians to meet and play together. The Swiss government has also advocated for peace in the Korean Peninsula for many years now. Having Swiss personnel stationed at Panmunjom as part of the Neutral Nations Supervisory Commission in Korea since 1953 is also a statement for peace. Perhaps an annual One Orchestra concert in Switzerland might indicate to the world that communication and cooperation between the two countries is possible, and fruitful.

Hyung Joon Won is the founder and executive director of the Lindenbaum Music Festival. His vision is to create the "One Orchestra" project that brings together young musicians together from North and South Korea. The Lindenbaum Orchestra first performed in 2009 under the direction of the world-famous Swiss conductor Charles Dutoit (current Principal Conductor of the Royal Philharmonic Orchestra). On February 27th 2013, Hyung Joon Won was invited to the Oxford Union to deliver a speech on Orchestra Diplomacy on the Korean Peninsula. An important step to achieving the One Orchestra project came on October 1st 2013 when he performed at a concert held in the Panmunjom Joint Security Area for the Neutral Nations Supervisory Commission.

Connecting Art and Science

An Interview with Jill Scott

Interviews with Innovators: You are the founder of the "Artists-in-labs" program at the Zürich University of the Arts. What was the biggest challenge you faced in setting up the program?

Jill Scott: Actually, there were three equal challenges:

• To find science lab directors who are interested in participating in such a program and can see the value of contemporary art as an interpretive platform for scientific inspiration.

• To build a core group of artists who are interested to learn about the teamwork of scientific production and can propose complimentary alternatives, inquires, and interactions.

• To secure the funding from organizations that value the potentials of know-how transfer and the creation of new cultural interfaces based on "hands-on" experience in laboratory environments.

IWI: How has the program evolved, and what plans are being made to keep it going in the future?

JS: Over ten years, the program has supported the placement of thirty artists into residencies based in academic science labs. So far, a wide variety of disciplines have been explored, such as in the life sciences, physics, engineering, and computing. The majority of these artists have been Swiss or Swiss residents, with the residency award given each year to artists for a period of nine months with funding from the Swiss Arts Council (Sitemapping BAK).

Over the last four years, this has evolved into the addition of cultural exchanges with India and China, whereby placements of Indian and Chinese artists into Swiss labs and Swiss artists into Chinese labs were developed in collaboration with Pro Helvetia. Last year we received a grant from the Agora division at the Swiss National Science Foundation, whereby six artists are still working together with six scientists to produce public debates and presentations. Consequently, we are diversifying and evolving, but the core aim remains the same: to give artists the opportunity to collaborate with scientists in various contexts and in a variety of ways.

IWI: Much of what is produced by the artists in the AIL program addresses the theme of convergence. How do you think transdisciplinarity manifests, and how can it be practiced when artists pursue research in a scientific lab?

JS: Well of course this is a big question, but convergence is not the only strategy under which art and science transdisciplinary practice can emerge. There are also other strategies like provocation, local intervention, awareness raising, education, or the co-creation of viable alternatives that attempt to address society's problems. We actually think that artistic research still has a long way to go before it evolves into its own more solid series of contextual debates, and we hope we are providing at least one interesting pathway for this to happen.

We have found that the one cannot easily compare what happens when, say, a text-oriented artist goes into a genetics lab that is focused on sleep, compared to a sound artist who grapples with the theoretical analysis of particle physics. The know-how transfer is different in each case. However, we do watch the process of exchange unfold and construct general recommendations for collaborators. For example, the need to first learn each other's vocabulary so that the conversations can run more smoothly is very important. Out of observations like these, we have endeavored to make the residencies longer and facilitate more discourse at later stages between the artists who are in residence. We also ask the artists to create a diary of their experiences, and ask the scientists to reflect on the particular contemporary art processes that they have encountered over the nine months.

IWI: You have been exploring the concept of neuromedia for a number of years. What works have you produced as part of this research?

JS: Neuromedia has emerged out of my own learning experiences in residencies and through studying Scanning Electron Microscope tissue samples at the University of Zürich over the last ten years. These residencies were not funded by the AIL program; they were supported through independent sources. A description of the works and the relevant science labs follows here: 1) Residencies in neural biology at the Institute of Molecular Life Sciences inspired SOMABOOK, an interactive media sculpture about neural development. Here I studied the development of the nervous system and the use of chicken embryos to understand the molecular basis of axon guidance and neural development. During development, axons (information transmitters) have to find their way to target cells to form neural circuits, the building blocks of the nervous system. The viewer in SOMABOOK can interactively combine audio-visual metaphors and manipulate interpretations of a dancer with scientific data about neural circuit formation, all accessible in a scaled up model of the spinal nerve. 2) THE ELECTRIC RETINA is an interactive sculpture about visual perception that explores the function of the visual system in zebra fish larvae, which have startlingly similar eye problems to human subjects. THE ELECTRIC RETINA is based on the rods and cones in the retina, and the viewer can compare interpretive underwater film-loops that depict visual impairment with basic neuroscience research on the visual system. 3) ESKIN started out as collaboration between Daniel Bisig and myself at the Artificial Intelligence Lab at the University of Zürich. ESKIN evolved through various stages to become an interactive project about tactile perception. It investigated how environmental perception relies on tactile as well as visual stimuli, the augmentation of sensory input, and the application of new somatic systems for the visually impaired. Because Artificial Intelligence uses a "build-it-in-order-to-understand-it" approach, ESKIN is the corollary work in progress germinated in the AI lab, exploring the potentials of interactive tactile interfaces and sound for robotics. In one manifestation, the viewer can investigate the sensory modalities of temperature, pressure, proprioception, and vibration to manipulate interpretations by a dancer of myths about the creation of skin-like landscapes. 4) DERMALAND is also an interactive project about skin, but it explores issues around embodiment in our environment. Our perception of our physical environment is changing due to the increasing impacts of global warming. For this project, I undertook a residency in Dermatological research at the University Hospital Zürich, and combined this with another in environmental science at the Forest and Snow Research Centre (also in Zürich). Here they are interested in stimulating public discussion around the burgeoning, damaging impact of UV rays on our soil. I constructed a large model of the skin and matched it to a Google Map of a part of the Kakadu National Park in Northern Australia. By using a Wii camera and magnifying glasses, the viewers can roam over the landscape and augment spots of reality where evidence of UV damage both on the skin and the soil can be witnessed. 5) I recently finished a residency in SymbioticA in Perth (Australia), where I was advised by experts in auditory research. This cumulated in a sound art sculpture called AURALROOTS, a project exploring a combination of tactile and auditory sensory perception. Inspired by learning about the tiny hairs on the auditory cells of the inner ear (called stereocilia), it consists of two hanging sculptures with related sounds and graphics made accessible on a touch screen.

IWI: Where do you see some of the most fruitful areas of shared interest emerging between neuroscience and contemporary artistic practices more widely at the moment?

JS: Well, I think that perception lies at the heart of both art and neuroscience research. There is a great history of interest in perception from all fields of the arts, and so artists are naturally drawn into this field of research and the current debates about cognition. I have been more interested in the hard wiring aspects of neuroscience, but there are many media and sound artists who are interested in exploring sensory perception in the cognitive sciences, as can be witnessed by the growing interest in cross-modal interaction, augmented reality, embodiment, or even the neuropsychological implications of social media. For example, I could point to Andrew Carne, Jane Prophet, or the artists in the show "Sleuthing the Mind" at the Pratt Manhattan Gallery in New York, in which I also recently participated.

IWI: With artworks that explore the connections between art and science, there will always be audiences who belong more firmly to one tradition or the other. Where do some of the challenges lie in creating artworks that occupy this space?

JS: Yes, I find that each person sees the work differently based on their own background and visceral reaction, but from the perspective of in cognitive science this would be a given. I am personally very interested to try and bring art and science audiences together, which will require an increase in know-how transfer. Currently, this is a frustrating endeavor because scientists tend to see my work as "not didactic enough", whilst artists see my work as being "too didactic". However, I still try to bring these different audiences together by inviting the scientists into my studio and presenting my work in both art and science contexts. As an art researcher, the collection and analysis of audience reactions and interactions are even part of my research. I also believe in the encouragement of transdisciplinary teams where artists are invited to play a major role.

IWI: Some say that science-based artworks are can be difficult for regular museum visitors to understand. Do you feel that more effort needs to be made in terms of how these works are mediated?

JS: Yes, I feel that more effort has to be made. I would like to see more artworks displayed along science projects so that the predominantly bottom-up, visceral, and experiential interpretations of artistic work can be compared to the predominantly top-down, calculated visualizations of factual science. I think that the Welcome Trust in London might be a good model to observe in this light. It may also help if artists showed more documentation of their own processes of production alongside their final outcomes.

IWI: Do you think that, in the new millennium, the arts and sciences are going to become more creatively connected?

JS: Yes, but it will take a while to achieve this goal. On the practical side of things, creativity in science tends to be very much linked to the building of experiments in order to understand the world; in the arts, in contrast, creativity is linked more to the building of experiments that create a post-reflective space for thought to emerge. It is through building of experiments that collaborative teams from both fields may have the potential to tackle some of the problems of living in the Anthropocene. On the theoretical level, creativity seems to be a an important common ground at the moment. This may be due to European philosophers like Latour and Stangers, projectionists who are aware of scientific discourse and the problematic ideologies of "progress". We are noticing in the AIL program that many people – from across different fields of the arts and the sciences – are becoming more interested in how you help others "to think creatively" rather than just telling them "what to think"! This is an encouraging trajectory for bridge-building.

Jill Scott is Professor for Art and Science Research at the Institute of Cultural Studies in the Arts, Zürich University of the Arts, Switzerland. She is the founder of the Artists-in-Labs Program, which places artists into labs studying physics, computer science, engineering, and the life sciences to learn about scientific research and explore creative interpretations of scientific practice. She is also Vice Director of the Z-Node PHD program on art and science at the University of Plymouth (UK), a program with 18 international research candidates. Her artwork spans 38 years of production, exploring themes about the human body, behavior, and body politics. In the last 10 years, she has focused on the construction of interactive mediated sculptures based on the research she has conducted in collaboration with neuroscience labs at the University of Zürich. This has included the study of artificial intelligent skin at the Artificial Intelligence Lab, human eye disease and cognitive interaction at the Neurobiology Lab, nerve damage in relation to UV radiation at the Dermatology Lab, and the development of neural networks in the pre-natal stage at The Institute of Molecular Life Sciences.

Creative Collisions: Arts and Science

An Interview with Ariane Koek

Interviews with Innovators: CERN is one of the world's largest and most respected centers for scientific research. Could you tell us in a few words about CERN, and about your role in leading its International Arts Development program?

Ariane Koek: CERN stands for the European Organization for Nuclear Research. It is a collaboration of 680 different scientific institutions from over 100 countries around the world dedicated to probing the fundamental structure of the universe. They use the world's largest and most complex scientific instruments to study the basic constituents of matter – the fundamental particles. In the Large Hadron Collider, the particles are made to collide together at close to the speed of light. The process gives the physicists clues about how the particles interact and provides insights into the fundamental laws of nature.

I am a producer, curator, communications expert, cultural strategist, and writer. As a producer (I was at the BBC for sixteen years as an award winning producer in both radio and television) I make things happen – putting together ideas, people, funding, and partnerships with a strong, direct focus. I am also used to managing the creative process, working with thinkers and artists – I did that at the BBC, for example. I was well known for working across science, the arts, politics, and history, for working with people at the cutting edge of new knowledge. I won a prize for science journalism, ran the first breast cancer public broadcasting campaign (which led to the formation of a new charity and understanding of breast cancer), as well as a national award for creativity. I am used to working with scientists and their ideas, as much as I am used to working with artists.

As a cultural strategist (I have worked as CEO of a major writers organization, and as a consultant for European organizations and festivals in helping them position themselves strategically), I am used to working with the mission of an organization, assessing its weaknesses and strengths, and then tailor-making a strategy for engaging with the arts which will fulfill the mission and take it to new dimensions. So far, for example, I have carried out a four month feasibility study for the arts program at CERN to see if it was needed, and, if so, how it could contribute to the organization as a whole and create added value. I created Arts@CERN to meet these goals, looking very closely at the people, place, and culture to create the arts program. So you can see how that ties in.

IWI: We know that collisions at CERN are very well planned. When things collide, however, they can fuse in strange and wonderful ways. How have the arts and sciences fused at CERN?

AK: My use of the words Colliding and Creative Collisions is very deliberate. I prefer to use the word collision over fusion to describe the creative process of art and science encountering each other. Not everything which collides leads to fusion, especially in energetic collisions. For example, when matter meets antimatter they annihilate, which in simple terms is one of the processes in which our universe was created. And so the process which we are doing with the arts and science at CERN is deliberately open-ended and described as a collision. Thus, the terminology and the direction of the program reflects the science as well as my belief in the creative process being open to anything happening, although of course there is an element of prediction as to what might happen.

IWI: I am sure you agree that scientists and artists are in someway equally creative, and that they both engage in the process of discovery. Nabokov spoke of art with "facts" and science with "fancy". In your opinion, is creativity and curiosity what scientists and artists share?

AK: Absolutely. Both scientists and artists are explorers and discoverers, not only of our universe, but also our place in it. Curiosity and creativity are two of the drivers for their explorations, as is the human imagination – the ability to think beyond boundaries and the known to go into the unknown. This is a particular passion of mine: the power of the human imagination and how the right creative conditions can be set up for it to flourish.

IWI: And are these the characteristics that bring about an attraction for scientists to the arts and the artists to science?

AK: Absolutely!

IWI: The arts program at CERN was designed to bring some of the most talented artists into contact with brilliant researchers and their work. How does the program curate these interactions, and why do artists seek them out?

AK: The ideas of science at CERN provide fundamental research material which acts like springboards for the imagination. For example, the Collide@CERN artists' residency program provides up to three months residency at the laboratory with curated experiences between the artists and the scientists. The new Accelerate@CERN artists' research program provides one month curated experience and the Visiting Artists one day program is even curated too, to ensure maximum impact. A recent Visiting Artist was Anselm Kiefer, and his experience was totally curated to match his interests, but also widen them further. This form of curation – where the curator acts as a guide or bridge between the two worlds, and works directly with the artist to help develop their thinking, working, and practice – is a fundamental aspect of the program; it is how I work.

Time and space give an artist the necessary elements to carry out this fundamental research, to explore and go into the unknown world of particle physics with the help of a curator – holding, invisibly, the process of discovery and exploration – by their side. Our first choreographer in residence Gilles Jobin said: "With Collide@CERN I learnt to learn again."

In our product- and deadline-driven world, artists have little opportunity these days to freely explore and let their imaginations go. They are bound by funding forms, timescales, and rarely have the opportunities to reflect, renew, and refresh. Our current Colldie@CERN artist, the data artist Ryoji Ikeda, describes his residency as a creative oasis that provides him with a once-in-a-lifetime chance. For fifteen years he has worked without a break. The residency has provided him with the time and space to fundamentally look at his own practice, work, and career. It has also helped to inspire future work.

IWI: Finally, and with a broader sense of the arts program in place, what are some of CERN's activities that have inspired your artists in residence to create new work?

AK: Everything from fundamental ideas around gravity, particle behaviors, and forces (ideas that have informed, for example, the dance piece QUANTUM by Gilles Jobin, our first Collide@CERN Geneva choreographer in residence) to the engineering beauty of the detectors themselves (as explored by the film maker Jan Peters, a work in progress which will be shown next year). We have welcomed musicians, visual artists, sound artists, filmmakers, fashion designers, electronica artists, dancers, choreographers, digital artists – the list goes on. Because CERN deals with the universe and how it came into being, it addresses fundamental questions which all of us as humans want to get to the bottom of! What we are looking for from artists is curiosity and an outstanding standard of work which matches the quality of our scientists – hence the title of CERN's Cultural Policy for Engaging with the Arts: Great Arts for Great Science.

Ariane Koek is an award winning producer, cultural strategist, and director who initiated, curated, and directed the Arts@CERN program between 2010 and 2015. Since January 2015, she has been a member of the CERN Cultural Board which oversees the program, and is now an External Expert for the European Commission on Digital Culture. Ariane led the International Arts program at CERN from Spring 2010, creating the laboratory's first arts policy – Great Arts for Great Science. Through the Arts@CERN program, three different strands for artist engagement were set up: the annual Collide@CERN artists residency awards; the new Accelerate@CERN artists research awards; and the Visiting Artists program, which has included such luminaries as Anselm Kiefer, William Forsythe, Esa Pekka Salonen, and Pipilotti Rist. Prior to CERN, Ariane had an extensive, award winning career at the BBC. Here she was a producer in both television and radio, working across the arts, sciences, the history of ideas, and politics. Following her work at the BBC, she became the CEO of the Arvon Foundation for Creative Writing, which ran 120 residencies at 4 different locations a year. Her career is dedicated to the belief in the power of the imagination to create personal, cultural and societal change.

 **ON SOCIETY**

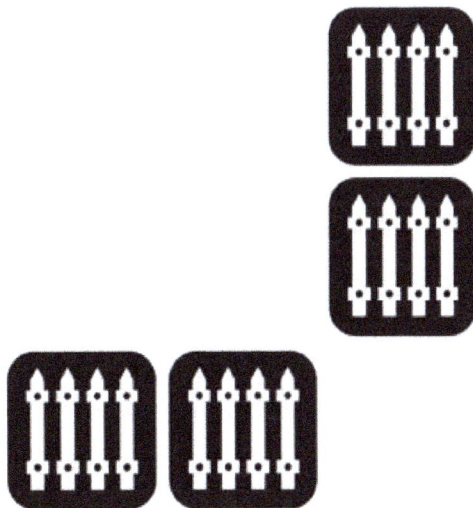

"Crisis" can be seen as a failure – on the part of those with power – to respond adequately and promptly to change. When maintaining the status quo is no longer an option, the need to intervene, to innovate, and to act comes into sharper focus. Today's crisis is one in which continuous technological, social and cultural acceleration risks disenfranchising a growing number of people. To counter this danger, many see the need for a more sustainable society wrought through a closer integration between social, political, and economic action. Questions being asked include the following: How can we locate new points of convergence and identify the opportunities they present? How can grassroots activities and local economies be better supported? How can we model cross-cultural interactions in the hope of developing socially responsible business practices and economically viable social ventures?

RAPHAEL H COHEN

Winning Opportunities

An Interview with Raphael H Cohen

Interviews with Innovators: You are a successful business angel, Professor, and CEO of your own company – an expert in promoting business innovation. How did you start on this path, and where would you suggest others begin?

Raphael H. Cohen: It began when my father challenged me to start a new division in his company. This is an unusual scenario that cannot be a model for others. What I have learned is that there are opportunities everywhere, and that most people do not see them or do not know how to seize them. After many years of teaching people how to identify opportunities and how to exploit them, I am now convinced that anyone with an open mind and the appropriate toolbox can create his job and jobs for others. Experience has shown that learning the tools, as explained in my book "Winning Opportunities", is enough to wake up those who are mentally agile. There is no need to create a start-up because behaving in an entrepreneurial manner inside an established company creates huge opportunities to boost one's career.

IWI: In your work, you have defined "opportunity" as the existence of an innovative solution responding to a market pain, need, or desire. What if those opportunities that arise also raise ethical concerns? How can such a dilemma be resolved?

RHC: This is a critical question. I was the first to introduce a course on business ethics at the University of Geneva in 2001 because I strongly believe that leaders have a responsibility to do the "right thing". This requires a clear awareness and understanding that whatever they do must be governed by an explicit set of values and governance rules. These values and governance principles should then act as filters to handle the dilemma you are referring to.

IWI: We are seeing a growth in the number of "hybrid" organizations, those that aim to generate income but also contribute socially. Should these two aims be considered independent of each other?

RHC: Any organization must in reality arbitrate four fundamental objectives: generate income and/or profit, ensure its sustainability, grow, and contribute socially. What you call "hybrid organizations" have the same four objectives, but they simply put more emphasis on the fourth objective: social contribution. If "traditional" organizations focus more on social contribution and less on generating profit they would become "hybrid". The question with this terminology is to define where "hybrid" starts... I personally do not like the idea that some companies make money and others contribute socially. I would rather merge the two and state that it is the duty of ANY organization to also contribute socially. This issue should be disconnected from the organizational structure. Hybrid and non-hybrid organizations can be top-down or bottom-up. Their governance is independent of their goals.

IWI: Many culturally or socially driven projects are seen to fail because key measures for evaluating their success are too revenue-oriented. What criteria should be used to evaluate those projects less geared towards "return on investment"?

RHC: Since resources are by definition limited, we have a duty to optimize their use. Launching the wrong project or a project that will not deliver the expected outcome translates into a waste of those resources. People who launch projects should thus be accountable for the proper use of the resources that they have received. The real question is to be clear on what measurable outcome is expected for each project. This is what I call the "Definition of success". It can be revenue-oriented but it is certainly not the only option. What is a must is to clearly express, before allocating and using resources, the measurable outcome of any project, even for a socially oriented project. Without this, it is not possible to hold people accountable for delivering this outcome with the minimum resources.

IWI: "Hybrid" organizations are often exposed to higher risk, financial instability, and may lack a clear business model. How can the credibility of such organizations be improved to help build new partnerships?

RHC: Credibility and sustainability will be improved by balancing the four objectives I have described. It is not enough to be socially oriented. Hybrids must also include the objectives of generating income, developing a business model that ensures their survival, and growing to increase their impact. Also, just because many hybrids generate intangible outcomes, doesn't mean that they cannot be measured as part of this process. Pain is intangible, and hospitals measure it. Customer satisfaction is intangible, but it is measured all the time. Social organizations can also measure their outcomes and build this information into a clear business plan.

If NGOs and social organizations want to increase their impact they should partner with other players by finding win/win approaches. This can only be achieved by understanding the world of those players. The effect of keeping an iron wall or a hostile attitude towards the private sector is that it will only prevent NGOs from maximizing their goals by leveraging such partnerships. Any such partnership requires a clear agreement on the outcome of the venture if it is to succeed. If they cannot agree they should not join forces because conflicts will become unavoidable. Once the measurable outcome has been defined, the next thing that should be addressed by the Stakeholders is the governance of their collaboration. Agreeing on the outcome and the governing rules is the best approach to avoid or at least reduce conflicts.

IWI: Last but not least, let's ask the million dollar question: How can artists and/or cultural organizations become more business-like, and would *there be guideline*s for them to make their creative endeavors economically viable?

RHC: Since any artistic or cultural project remains a "project", there are certain rules that apply. It happens that the "business" world has studied those rules to optimize a project's outcome. If artists or cultural players learn those rules they will be able to apply them to their projects and their relationships with other players. This should significantly increase their chances of success. I am a strong believer in the power of bridging different worlds.

Raphael H. Cohen is a professor, lecturer, author, serial entrepreneur and business angel. In his academic work, he is the director of the entrepreneurship & business development specialization of the eMBA at the University of Geneva. He has also been the MBA European Academic Program Director of the Thunderbird School of Global Management, and director of the first course of entrepreneurship at the Ecole Polytechnique Fédérale de Lausanne (EPFL). In parallel to his academic work, Raphael has been managing an international group of companies since before even earning his PhD in economics in 1982, and is currently the owner and managing director of Getratex SA. He sits on the board of several companies in different countries, including a bank. Through this extensive experience of the business world and as an expert in agility and corporate innovation, Raphael offers executive education, mentoring and consulting services to senior executives, bankers, directors, middle management, and entrepreneurs. He has developed the IpOp Model which provides a toolbox to improve the outcome of innovation, corporate entrepreneurship and employee commitment. The model aims to help businesses identify opportunities that could bring them a competitive advantage, and minimize the waste of resources on unproductive projects. In addition he teaches and promotes another toolbox for "just" / "caring" leadership, which aims to deliver true employee engagement.

Approaches to Responsible Behavior

An Interview with Susan Schneider

Interviews with Innovators: You are a psychologist with a research interest in organizational behavior and cross-cultural management. What aspects of management culture are of particular interest to the field of psychology, and why?

Susan Schneider: Management is about managing people and managing teams. Understanding what motivates people, how leadership roles are formed, and how group dynamics operate are core interests in psychology. There are many ways in which we can begin to address these questions. For example:

• What motivates people? – Are people motivated by tangible rewards (such as money), status, or by opportunities to develop and contribute to society?

• How are people motivated? – Is it through satisfying these needs, by managing expectations of how their efforts will lead to performance and rewards that they consider valuable (expectancy theory), or by comparing their efforts and outcomes to those of others (equity theory)?

• What personality characteristics are important for effective leadership? – Is it more important that a leader should be charismatic or authentic in his/her actions?

• What behaviors make leaders effective? – Is it through promising rewards for performance (transactional) or inspiring people through their actions (inspirational)?

• How do groups function best? – Can we, for example, understand the norms that govern behavior (such as to compete or cooperate) and how group decision making compares to the decision-making processes of individuals?

Most recently our research has sought to understand why certain people in organizations engage in socially responsible behavior. We defined social consciousness as comprised of cognition (moral reasoning), emotions (positive and negative affect), and values. All of these constructs are derived from psychology research. As part of the RESPONSE project we have explored how different patterns of socially responsible behavior emerge in different decision making scenarios.

IWI: What is social responsibility and how does it lead to social entrepreneurship?

SS: Social responsibility has most often been looked at from an organizational perspective, in other words **corporate social responsibility** (CSR). Although many different interpretations are available, CSR can be understood as "the voluntary integration by companies of social and environmental concerns in their commercial activities and their relations with their stakeholders" (European Commission). Issues faced include those around discrimination, corruption, environmental footprint, sustainable sourcing, and so on. It offers a top down perspective on social responsibility. But, as mentioned earlier, another question can be raised: What makes some managers more socially responsible than others, i.e. to take decisions for the benefit of society? And how can we help to develop more socially responsible managers. Again, this is most often addressed in the context of the organization. Social entrepreneurship is more often driven by those individuals who have opted out of organizations. Social entrepreneurs' main reason for being is to address social issues and to make profit to serve that purpose. For most companies, profit is the primary purpose, with social benefit taking a secondary role. In our research we found that most managers in companies considered social responsibility so to avoid harm. In contrast, the main concern for social entrepreneurs, I think, is to do good. Whilst managers tend to limit their perspective on social responsibility within the domain of a company's products and services or corporate image, social entrepreneurs try to address difficult social issues, such as poverty or healthcare, through stimulating new forms of action.

IWI: What are a few of the most important challenges and opportunities in regard to social entrepreneurship?

SS: The most important challenges are to be able to address social issues and remain viable as an organization. That means having a workable business plan that can be sustained. There are many opportunities to find causes to invest in, but good management is a necessary ingredient! Even companies that have tried to address social issues, for example P&G tablets for clean water, found that it was difficult to create a business model to sustain it. Problems also developed with regard to microfinance, especially when commercial banks saw it as a business opportunity. There seems to be different logics operating between the public and private sectors which makes it hard to bring them together. The new trend in developing public private partnerships (PPP), such as bringing together UNICEF and P&G, tries to bridge this gap.

IWI: How is corporate philanthropy shaping the identities of corporations?

SS: Corporate philanthropy is nice to have but it is not enough. It is almost too easy to give money and to say look at us! For this reason it is often dismissed as public relations efforts and does not really reshape corporate identity. Corporate responsibility is about "who we are" not "what we do". Many companies try to address social issues using their core competencies, for example IBM providing computers to inner-city schools. While this may be considered philanthropy, it is also intended to help develop the next generation's computer literacy, which means future potential customers and employees. This is a win-win situation, or "creating shared value" according to Professor Michael Porter of Harvard.

IWI: What is the balance between social and financial performance in the corporate world?

SS: There are now many studies which show that corporate financial performance (CFP) and social performance (CSP) is not a trade-off, but can be mutually reinforcing. Clearly, companies are not charities. They need to make a profit to exist and to continue to provide products/services as well as jobs. This was the definition of CSR by the famous economist Milton Freidman. For him, addressing social issues was the role of government and using company money towards this end was "highway robbery." But companies also need to consider the needs of stakeholders, and not just stockholders. Profit at any cost is not acceptable. Big bonuses and big layoffs are no longer tolerated, hence the recent Occupy Wall Street movement.

Furthermore, and perhaps most importantly, the results of companies rated as socially responsible tracked on FTSE 4 good or the Dow Jones Sustainable Index (DJSI) and other indexes show similar returns over time. However, the debate continues as to whether social responsibility can be regulated or if there is a "market for virtue", which Professor David Vogel of U C Berkeley doubts. In other words, can consumers and investors really insist on social responsibility? There is now a major increase of socially responsible funds being created.

IWI: How is it possible to train or teach social responsibility, and to motivate responsible behavior in individuals as well as corporations?

SS: Training. As part of the RESPONSE project conducted by a consortium of business schools, we tested different approaches to developing socially responsible behavior in managers. We were interested in trying to understand how socially responsible behavior might vary across cultural, legal, industry and organizational contexts, and identify factors that explain such behaviors at an individual as well as organizational level. One hundred managers in 4 companies took part in the experiment. Some managers had a full day course (executive education) on CSR. Others followed 6 weeks of either relaxation or meditation training. Participants filled out questionnaires before and after that assessed changes in their values, emotional states, and how they respond to different scenarios around decision making and moral reasoning.

We found that social consciousness can be developed through approaches based on relaxation and meditation, whereas cognitive approaches (executive education) may be limited in its impact. To our surprise, the executive training seemed to have a somewhat negative impact on certain areas of decision making, while relaxation and meditation training both had a positive impact, but in different ways. Relaxation training had more impact on values (promoting concern for social justice and protecting the environment for example) while mediation training had more impact on emotions (promoting happiness, inspiration and courage). We refer to this impact on decision making as reflective versus reflexive, or based on moral reasoning versus moral intuition. We're now working to understand what implications this research has for integrating social responsibility at the individual and organizational level. The aim is to embed social responsibility in identity, i.e. to drive responsible behavior not because it is "what we do", but because it captures "who we are".

IWI: What role can education play in moving from ideas of social responsibility to driving a new generation of socially responsible entrepreneurs?

The role of education is absolutely critical. We set up a Masters in Management program in 2005 with a course on Business in Society built at its center. Based on a multi-disciplinary approach (accounting, finance, marketing, IT, HR etc...), the course brings in representatives from large corporations and NGOs to discuss social responsibility. For example, representatives have been brought in from Procter and Gamble to discuss marketing and sustainable development, from Shell to discuss global diversity, and from Ethos to discuss social investment. A key part of the course is that students conduct projects aimed at "doing something for the community". Here students learn skills in project management (logistics, organization, planning), fundraising (finding sponsors and approaching donors), communication (publicity and social media), and collaboration, but also get first-hand experience in seeing how projects can make a difference and the challenges faced in making them sustainable. We've learned that social entrepreneurship can start early and that young people really want to be involved. Education at many different levels can provide the spaces and opportunities to make this possible, but we need more champions and both organizational and institutional support (such as accreditation agencies). The benefits of – and pressure for – socially responsible behavior within companies is increasingly apparent; we need to respond to that call.

Susan C. Schneider is Professor Emeritus at the University of Geneva, Switzerland. Her teaching activities include topics in organizational behavior and cross-cultural management. In 1997 she published Managing Across Cultures with Jean Louis Barsoux (now in its third edition with Günter Stahl, published by Pearson). It has been translated into Chinese, French and Dutch. In addition to her academic work, Dr.Schneider has taught seminars for multinational companies and cultural organizations. Her research activities focus on cross-cultural management, diversity and social responsibility. She has published articles in Academy of Management Review, Strategic Management Journal, Human Relations, Organization Studies, Human Resource Management, Human Systems Management, and Political Psychology.

Driving the World with Social Innovation

An Interview with Byungtae Lee

Interviews with Innovators: You are the Director of the SK center for Social Entrepreneurship. Can you tell us about how the collaboration between KAIST Business School and the SK center is leading to new model of education geared towards the promotion of social enterprise?

Byungtae Lee: Although South Korea has many social problems common to the rest of the world, it also faces its own unique challenges. As South Korea has transformed from an agricultural economy through an industrial economy to an information society in a very short period of time, it faces many new problems, including the creation of under-privileged and vulnerable groups. These include an aging population living in poverty, exiles from North Korea, multi-ethic families, the homeless, an increasing population with mental health problems, unemployed youngsters, and so on. After the foreign currency crisis of the late 1990s and the recent global financial crisis, social enterprise has been recognized as offering possible solutions that are an alternative to free market or government welfare programs. In the last few years, through programs promoted by South Korean central and municipal governments, many social enterprises have been founded and tested. While public programs and policies have been very successful in driving social enterprise and raising public awareness for this new kind of business organization, many of them failed to develop sustainable business models; the majority of them barely survived on government subsidies.

As the chairman of SK Group, Mr. Choi has been a champion in promoting social economy and enterprise. Although the SK Group, under his leadership, have experimented with diverse programs for encouraging social entrepreneurship, they soon recognized that social enterprises can only be viable if they develop sustainable business models; this requires competent leadership. KAIST Business School, too, has been searching to develop suitable business models of this type. The strategic choice for KAIST was to focus on fundamental issues faced by humanity; social and environmental sustainability are key challenges for us all. This shared vision for training a new generation of business leader – with a strong orientation towards creating shared value for a better society – led to a collaboration between these organizations to create new model of business education.

Under this vision, our collaboration has produced three entities: the Social Entrepreneurship MBA program, the SK Center for Social Entrepreneurship, and the KAIST Venture Investment Holdings. Our SE MBA Program is the only full time MBA program in which students are required to create social enterprises during, or immediately following, completion of their training. All students are fully supported financially by a gift from SK Group's Happiness Foundation. The SK Center for Social Entrepreneurship has two primary goals: Developing the SE MBA Curriculum and providing incubation and acceleration support for social enterprises. Most MBA programs in South Korea are a copycat of American MBA programs. The new SE MBA program, in contrast, has been created from scratch. As a result, we are continuing to experiment with and refine the main curricula and co-curricular programs. In the program we have sought to harness both the strong theoretical work emerging from the KAIST Business School and the practical expertise and leadership skills found in the SK Group. The third entity, KAIST Venture Investment Holdings, is a venture capital fund founded through a generous personal gift from Mr. Choi, the chairman of SK Group. This venture capital firm provides seed funding to social innovators. The lack of capital available to young entrepreneurs is often cited as a critical barrier to innovation in South Korea. This is particularly the case when considering the social enterprises that seek funding from capital market.

In addition to these three core areas of attention, we have also helped other schools to create similar education programs through sharing our knowledge and lessons learned. We sense strongly that we have raised the bar for social entrepreneurship in South Korea through attracting and training talented young entrepreneurs.

IWI: You have discussed three terms which capture the process of becoming a social entrepreneur: Being, Knowing, and Doing. What is the significance of these terms?

BL: The three terms – Being, Knowing, and Doing – originated with the book, "Rethinking the MBA: Business Education at a Crossroads" authored by Srikant M. Datar, David A. Garvin, and Patrick Cullen. They criticized business education for having failed to find a balance between these three components. In other words, MBA education teaches only knowledge (Knowing) but has ignored value systems (Being) and practical leadership (Doing). We viewed our Social Entrepreneurship MBA curriculum as a bold innovation in business education aimed at fulfilling all of these three essential components. In particular, a value system oriented towards a social mission is the foundation stone of any social enterprise. Many courses and extra-curricular activities that come under the umbrella of "being a social entrepreneur" are designed to identify social problems and find innovative solutions. The second component – Knowing – is based on core management courses, and the third component – Doing – is materialized through business planning and incubation programs. Experienced executives and entrepreneurs assist our MBA students as mentors. They are encouraged to implement their ideas as soon as possible through multiple rounds of market testing and an ongoing revision of their business models. This market-oriented business preparation sharpens their abilities in designing and implementing **social innovation**.

IWI: How is the KAIST business school tackling issues around social and environmental sustainability within an academic context?

BL: Whilst most business schools claim that issues of sustainability are addressed in their curricula, KAIST Business School is unique in the sense that we have two specialized programs focusing on this issue in particular. Its program involves not only discussing issues of social sustainability in the class room, but also creating businesses that can directly address those issues in the real world. Green Growth Management School (founded at the same time with SE MBA) provides education and research on environmental sustainability. Of course, these two programs are not typical business programs. When we introduced these two programs under my leadership as Dean, many faculty members and stakeholders were very reluctant to get involved. Building consensus amongst stakeholders was a considerable challenge, as was securing resources to implement the program itself.

Business schools generate the majority of their revenue through MBA programs (or undergraduate programs, as here in South Korea). Usually, MBA programs are one of the most expensive academic programs, along with those in law and medicine. Hence, MBA students are either sponsored by their employees or self-financed from savings after a few years of employment, and expect higher returns from MBA degrees. Students who are interested in Social Enterprise, however, are different. They cannot expect immediate returns from the SE MBA degree on the basis of creating social ventures, and, of course, there are not many social enterprises that can sponsor their employees' education. Therefore, we created this program with full financial support.

When setting up the program within an academic context we took a number of key steps. Firstly, the program is designed to focus on social issues first and foremost. Whilst curricula are often developed on the basis of an instructor's individual experience, we sought advice from external experts to design a program that could best address such issues. This included holding open workshops with students, external academics, global partners, and industry executives. Secondly, the program harnesses campus-wide intellectual expertise in order to develop the most innovative solutions. KAIST is a world leading university for science and technology. As the KAIST Business School pursues a program that aims to integrate management and technology, it is well integrated into the wider university, making this possible. Finally, the curriculum is monitored and assessed by the KAIST curriculum management system to ensure the quality and effectiveness of the program.

IWI: What is the purpose of tying an MBA program so closely to academic research, and what do you think is the proper role of industry in such a program?

BL: The reason why we need academic research on social innovation is that we still only have a few examples of successful social enterprises. Many social ventures are created with a value-sharing vision but without sufficient resources to support that vision. An orientation towards value-sharing, however, is often viewed as a constraint from the perspective of financial performance. That is why traditional financial markets don't actively invest in social innovation. The market for ventures that address social problems is also very small; another reason why conventional businesses avoid investing in them.

However, we observe that a strong social mission can be a strategic asset, not only in helping convince stakeholders but also to drive changes in consumer behavior. Such niche markets can also be profitable when they have a strong, local embeddedness. Through these two observations we see that social enterprises often reside at the borderline between the public sector and free markets. This position means that such enterprises need to pay more attention to non-market regulatory competition and support dialogue with the public sector.

For these reasons, managing social enterprises is not like managing conventional businesses. To understand these issues better – and find solutions – we need more rigorous academic research. Drawing from Industry, we have already experimented with (1) channeling Corporate Social Responsibility (CSR) efforts and resources into the social economy sector, (2) transferring management skill sets into MBA education (mostly through incubation programs), and (3) promoting awareness of social innovation practices.

IWI: Cultivating social entrepreneurship – bringing together people who have vision, passion and integrity – is no easy task. What has been your experience so far on that search for potential entrepreneurs, and how do you support them after the SE MBA program in finding real-world employment?

BL: We are competing with many other entrepreneurship programs, largely those interested in driving technological innovation. The current government in South Korea is particularly interested in pushing such programs, and they are making money easily available to the budding entrepreneur. Young innovators, however, seem to seek the Silicon-Valley-style success story, one in which an idea is quickly incubated and profit made. Under these conditions it has been much more difficult for us to attract talented innovators who are willing to commit to a two-year program. Until they experience management problems, however, they don't realize the importance of a solid management education. Equally, though, creating a business venture and completing an MBA education simultaneously is a big challenge for our students.

After completing the SE MBA, the KAIST Business School has a number of support programs in place to help students find employment, similar to many other global programs. In addition, our school is one of the premier business schools in South Korea, with a strong branding as an elite educational institute. Statistically our traditional programs claim 100% employment within three months of graduation. The SE MBA program is unique in producing job creators rather than job seekers; this is evidenced in that all of our graduates successfully launch social ventures either during, or upon completion of, the program.

IWI: Turning our attention to the bigger picture: In his book "Capitalism 4.0: The Birth of a New Economy in the Aftermath of Crisis", Anatole Kaletsky suggests that capitalism is "not a static set of institutions but an evo-

*lutionary system that reinvents and reinvigorates itself through crisis".
How do you think Capitalism is likely to transform in the coming decades?*

BL: I believe in the collective wisdom of mankind to solve social problems, but cannot predict the near-future evolutionary course of Capitalism. I can, however, strongly predict that social enterprises will become
meaningful experiments that shape its evolutionary trajectory. In fact, if
successful, there will be no distinction between social and conventional
enterprises in the future since conventional ones embrace the power of
value-sharing.

On the other hand, society will soon re-write social contracts by reflecting technological progress and productivity gains by smart machines. We
are suffering now because the globalization of business, rapid demo-
graphic change, and technological impact takes place at the same time,
creating radical disequilibria in the global economy and driving negative
social change. Capitalism will evolve along the course of new equilibria
once the dust of these current, radical changes settles.

*IWI: There are parts of the world that we term "hotbeds of innovation".
However, innovation can undermine the conditions from which it originally arose; growth in technology has resulted in "a race against the machine", apparent through the number of jobs being replaced by technology in an increasing number of sectors. Is technological innovation the
savior or the curse?*

BL: As an economist, I am not sure whether technologies destroy or create jobs on the whole. Often, jobs lost in the industrial economy are not
really lost but merely "relocated" to developing countries. What is clear
to me, however, is the rapidity with which jobs are changing. Given the
productivity-gain achieved through machines in developed countries, it
will not be easy to recover jobs if we keep our current system of employment. As productivity has increased, mankind has begun to work less
hours. In the early stage of industrialization we worked a lot more hours

than today. Hence, to create enough jobs, we have to re-write labor contracts. This is a matter of collective decision-making as to how we share economic outputs amongst ourselves. Hence, if we maintain the current system, it is unlikely that we will "beat the machine" and create jobs enough for everybody. If we try to change social contracts, however, it may be possible.

IWI: What role do you think social values play during a time of economic crisis and job market failure? How would creating sustainable social values through social entrepreneurship feedback into community experience and increase its impact in general?

BL: I view social enterprises as pursuing ideals. Our public welfare programs are based on the belief that we can build a better society by caring for the vulnerable. But, often, government programs exhibit inefficiencies and become buried in decision making around moral dilemmas. In contrast, private enterprise produces value through innovation and driving efficiencies. Hence, theoretically, social enterprises can aim to preserve the best of these two very different types of institution.

Of course being an idealist does not guarantee success; however, that shouldn't stop us from pursuing our ideals. I often find inspiration from the passion our younger generations exhibit for life and social values. Once I met the Dean of the Oxford Business School; he claimed that we, the older generation, underestimate the passion the millennial generation have for creating a better, healthier society. I totally agree with him.

Byungtae Lee is a professor at the KAIST College of Business (South Korea) and a director of SK Center for Social Entrepreneurship. Before joining KAIST, Byungtae Lee held the position of assistant professor at the University of Arizona, subsequently becoming an associate professor at the College of Business Administration at the University of Illinois at Chicago. He was the president of the Association of South Korean Business Schools, and a co-director of the Asian Affinity Group (AACSB). In addition to his active career in academia, Byungtae Lee has also considerable experience in the field. He was a General Manager, CIO at SindoRicoh, and advisory professor at Kookman Bank, Shinhan Bank, Samsung Life, KRX, etc... He earned a Ph.D. in Business Administration (with a major in Management Information Systems) from the University of Texas at Austin.

Making Impact First

An Interview with Chong Soo Lee

Interviews with Innovators: Models of social investment are emerging world-wide with the aim of driving positive social or environmental change. Why is social investment so important today?

Chong Soo Lee: Social disparities and socio-economic malaise are threatening our chances of building a sustainable future. We are facing such issues as low economic growth, an aging population, youth unemployment, and economic bipolarization. As these issues become more serious, the resources needed for dealing with them simply cannot keep pace. If we fail to properly manage the enormous challenges that we are all now confronted with, it seems hard to imagine how we can build a sustainable society at all.

Social investment and social enterprise are emerging as alternative approaches to traditional business models. They can simply be defined as "impact first" investment, meaning that it's primary purpose is to create social value. This doesn't mean that generating financial return isn't important, only that it is of secondary importance, with the range of acceptable rates lower than market return varying according to the investor in question. Social investment represents a paradigm change in how we resolve social issues faced in contemporary society, in part because it aims at generating self-sufficient ways of funding possible resolutions. We need to create best practices in these areas locally and work hard to promote them internationally.

IWI: What steps are being taken by South Korea to support this new movement?

Seoul Social Investment Fund (SSIF) is a representative case of the social investment movement. Established in 2012 by the Seoul Metropolitan

City Government under Mayor Wonsoon Park, and managed by Korea Social Investment, SSIF can be said to be pioneers of this approach in South Korea.

KSI's vision is to create a sustainable future through social finance, so developing a more inclusive and responsible society. Loans are offered to businesses that create value in social, environmental, cultural, and educational areas, in particular those expanding the base of social economy by supporting social businesses, social cooperatives, and social ventures. We see this as an effective way of attracting private resources for the funding of preventive interventions that treat social problems without being either short-sighted or resorting to post-crisis remedial social services. We hope to see this movement gain momentum and attract investors from the private sector with an interest in social causes.

IWI: What would be an example of the kind of business that Korea Social Investment supports?

CSL: A social business is one that can create socially oriented outputs. An ideal business model is one that is also able to make financial value in the process, which helps the business to build a more sustainable operation. Korea Social Investment has made loans available to businesses that they believe can generate great social value on an initial investment. This has included a number of large and small solar energy generating projects, a car sharing business, a social franchise restaurant business, and social housing projects.

For example, we have supported SOCAR, a car-sharing project that aims to reduce carbon emissions, save on parking lot usage, and save money for car owners. On becoming a member, a car can be rented at a low price and for a minimum of 30 minutes. Car return is through public parking lots and arranged through a smart phone application. SOCAR is now expanding its business from Seoul to other major cities and travel destinations in South Korea, including Busan, Jeju and Gangreung. Secondly, the project Solar Energy Plant (run by the social enterprise

Energy-Peace) is a scheme to install solar panels in urban environments, such as over water purification plants, subway stations, and even school rooftops. The company is currently running sixteen solar power plants, with the energy produced either made available to help energy-poor households or sold. Profits made are returned to the energy-vulnerable groups involved. Finally, the Social Housing project aims to finance social construction companies to build houses for low-income families, creating jobs for construction workers and providing more affordable housing. The scheme works through Seoul City inviting social enterprises to build, renovate and run housing for vulnerable groups with financial backing (such as low interest loans) provided by KS Investment. We are currently supporting three social enterprises and three housing coops in this endeavor.

IWI: How is it possible for social ventures to win over investors and develop sustainable relationships with stakeholders?

CSL: Many socially oriented ventures can only attract investment based on their financial potential, the robustness of their business models, and their exploration of innovative ideas; this is independent of any social value they may aim to create. In South Korea, a small number of social venture capitals have started to operate, and investments into promising social ventures have been made (such as Tree Planet, Green Car and Law & Company). However, most examples of social ventures have a weaker financial potential, so they need to prove the size of the social value they claim to be able to create in order to attract "impact first" investors and philanthropic investors who are willing to give up some degree of financial return in exchange for generating social value. The question then becomes how to prove that you are, in fact, creating social value. The development of a methodology for measuring social impact is absolutely essential; without one, it will be a struggle to attract impact investors from any part of society. At the initiation of Korea Social Investment, the Social Impact Evaluation Network (SIEN) was formed in order to promote the development of an impact measurement infrastructure in Korean society.

IWI: What roles do innovation and the convergence of ideas across different sectors have to play in developing new approaches to the social issues we currently face?

CSL: Many people believe that traditional approaches are not enough to tackle the social issues that we are confronted with. This is because social issues are not only very complicated but also extremely diverse in nature. This is why we need to consider new approaches around convergence and innovation in order to generate a more joined-up approach to these serious issues, to both increase and diversify the flow of investment, ideas and talent that can come together to tackle social problems.

Ideas around convergence and innovation have already become an important driving force in mainstream finance and business. Such trends have also started to influence the social sector as well. This phenomenon, even if it's only still at an early stage, can really be seen in the increasing number of social ventures active in South Korean society. Given the fact that South Korea is very strong in IT and mobile development, a number of social ventures are exploring new, innovative points of convergence between mobile technology and social business. For example, one venture is adopting mobile communication systems into smart phone applications for social workers to help them better work together with senior citizens under their care. Another is enabling the mobile-transfer of money through Social Networking Services so that small donations can be made to support social investment. In the case of Tree Planet, this mobile game for planting seeds and growing trees actually leads to the planting of real trees in decertified land.

I think an important aspect of driving these changes is to engage more in conversation across sectors. Conferences and workshops offer the opportunity to exchange knowledge and experience around innovation with experts from different cultural backgrounds. Such events can also offer the opportunity to discuss new approaches with clients who use, or are interested in using, hybrid models to tackle social issues.

Chong Soo Lee is one of the most renowned figures in the micro-credit bank and social investment field in South Korea. Early on in his early career, he gained experience in international trade and finance as a member of Chase Manhattan Bank, successfully establishing Asian branches of Westpac Bank and Advanced Bank of Asia. As a CEO of AON Korea he has led the insurance consulting field. He established Social Solidarity Bank, the largest micro-credit institution in Korea, in 2002. Since then, Chong Soo Lee has focused on building up virtuous circles between finance, social welfare, and individuals working towards creating a better society. He is also the founder and CEO of the Korea Social Investment, a corporation that was established with the purpose of introducing new social investment and social finance models to South Korea.

Hierarchy of Failure, and Solutions

An Interview with Raymond Saner

Interviews with Innovators: You have argued in your work that the social crises we face reveal four levels of systemic failure, which together undermine our capacity to build a healthy and sustainable society. What are these systemic failures, and how can they be corrected?

Raymond Saner: In my analysis of the current societal malaise, I identify four levels of systemic failure. "Market failures" can result from domination by one (monopoly) or two (duopoly) suppliers of goods or services. Hence, the buyer has no alternative than to buy from whatever the monopolist puts out in the market and at whatever price he wants. Market failure can also occur because there is no supplier in the market to meet needs, for instance, for public goods like security (e.g. street lighting, the police). There might also be incomplete markets, where markets fail to produce enough goods or services for all, like education and healthcare. Another big market failure is rooted in the failure to take externalities into account (e.g. the environmental pollution caused by industrial production which is not included in the price of industrial goods). Market failures can be corrected through dismantling monopolies, through regulating price fixing, through making government regulations more transparent, and by making it easy for the public to access government services, for instance through e-government-based service provisions.

"Government failure" occurs when governments fail to intervene to correct market failures. This can be due to corrupt practices by government officials who provide more favorable market conditions to one enterprise than to its competitors. For instance, such market distortions can be orchestrated by penalizing the competitors, making them pay additional fees, putting them through excessive audits and overcharging them through unfair taxes, or simply by making market sensitive government information not available to

157

all economic actors entitled to such information. The same can be observed in the case for social services when, for instance, some citizens are excluded from public services that should, by law, be made available to them (i.e. discrimination). Government failure could be corrected through new laws (in democracies), provided there is a judiciary which can be called upon to intervene in situations where laws are deliberately not implemented or not made public. Another way of correcting government failure is through citizen action.

The third form of failure is "academic failure", or lack of civic courage and leadership by university researchers and teachers who avoid presenting research findings that could put them into conflict with authorities or dominant groups in society. Academics are supposed to push boundaries, to contribute to new "discovery" and knowledge, and not to withhold what they have learned from their research, and hence not to obfuscate reality. Academic failure can arise when university professors are members of a political party that holds power, such that the publication of research that could be interpreted as critical of the current status quo is avoided or hidden through deliberately quasi-scientific language not easily understood by ordinary citizens. An example would be the hesitancy of Swiss social scientists to analyze the impact of the dominant market position held by a few media groups owning the majority of leading Swiss newspapers. Similarly, there is reluctance by some political scientists to analyze political party financing and the impact of lobby groups on democratic decision making processes in parliaments, be that at national or regional level.

The final level of the hierarchy of failures is "civic failure". This occurs when citizens are not enabled, nor have the practice, to articulate what they consider wrong and in need of corrective action. Citizens can, and should, convey their concerns to academic scholars, to members of parliaments, and to the owners and lead managers of businesses whenever they observe a major dysfunction that needs to be corrected but does not get remedied due to market, government, or academic failures. Social articulation of major dysfunctions can be reinforced through popular culture and mechanisms of civic participation. For instance, civic participation is possible in Switzerland through direct democracy (right to launch a referendum or initiative). Another example is the Stanford University-based liberation technology activists who use social media to correct government and academic failures.

IWI: How serious are the problems we face in society today, and is there more of an acute need for change today than in the past?

RS: To answer this question adequately, one has to move from a local to a global perspective and also step back from a Euro-North American myopic view to look at the totality of our planet. We should take note, from a global perspective, that the number of Least Developed Countries (LDC) has risen from 24 in 1971 to 48 today. A short comparison serves as an eye opener: Based on World Bank 2013 statistics, the LDC group consisting of 918.9 million people had a GNI per capita of USD 868.00, life expectancy of 61 years, and CO_2 emissions of 0.3 metric tons per capita. By contrast, the USA consisting of 319 million people had a GNI per capita of 53,740.00 USD, a life expectancy of 79 years, and CO_2 emissions of 17.6 metric tons per capita.

This contrast is extraordinary, and the main message would also hold if this were to include comparisons between the EU, Japan, Singapore and the LDCs (or the even larger group of low-income developing countries). This contrast highlights the difference between deep poverty and high wealth. There are, of course, also the super rich in poor countries and very poor people in the richest countries of the world. The point here is not about inequalities – that would be an obvious and easy point to make. The main point instead is to state that current global imbalances are not sustainable.

Continued large-scale imbalances of wealth and living standards result in increased migration, armed conflicts, crime, violence, and extremism of all sorts. Should the industrialized and emerging countries remain incapable and unwilling to stop climate warming, further environmental destruction appears inevitable, and with that will come an increasing vulnerability to large populations. This, in turn, will most certainly lead to more instability and more conflicts, not less. To cling to the mantra of "business as usual" is to deny that such imbalances will increase the environmental, social, and economic costs to mankind, and thereby endanger the survival of the majority of people. It is time to accept the fact that we are all inter-dependent, and that the notion that a country could be independent of the rest of the world is both a delusion and a form of eco-suicidal autism.

IWI: Failures at the level of markets, governments, academia, and civil society define the hierarchy of failure. What would a hierarchy of solutions look like, and who should take the lead in helping us develop a more sustainable society?

RS: Leading initiatives have already been started, and many are going in the right direction of helping our atomized societies to find ways of recreating social bonds and communal co-existence. We need to bring back to the awareness of our citizens that we all depend on being embedded in multi-fold interdependencies, and that we need to better share the limited resources required for our survival and the survival of the global village. People have lost the ability to reach out to others – to make contact – which means not to busy oneself only within a closed circles of similar types of people, as can often be the case with closed social media "friends of" networks. We should learn, once again, to learn with others outside of our habitual networks, which means to engage in critical discourse, to agree to disagree, to analyze issues, and attempt to find solutions to identified problems. This all implies critical thinking and tolerating potential dissonance and ambiguities until solutions become figural. Our business culture has unfortunately become myopic and too straitjacketed by inordinate efficiency mantras.

Excellent examples of international networks that foster the exploration and sharing of social innovations include Globelike or the initiative taken by the International Social Science Council (ISSC) to conduct transdisciplinary research into how global change could be fostered to ensure the implementation of what is termed the "triple sustainability bottom line". (This refers to the sustainable development of the social, economic and environmental sectors of our society captured in the Sustainable Development Goals (SDGs) currently being negotiated by the countries who are members of the United Nations). Another initiative is the search for cross-sector social partnerships that bring together state and non-state actors to create common projects between private sector companies and civil society organizations. The survival of our world requires collective actions and multiple levels of collaboration. This cross-boundary cooperation requires going back to the basics, i.e. trust, empathy, ethical conduct, and a constructive competition based on new ideas generating beneficial social impact. Citizens form different networks should tackle the four failures with a shared vision of an inclusive and dignified world for all.

IWI: You speak of sustainable development in terms of economic, environmental, and social development. What role does culture play, and how could a "culture of sustainability" be defined and practiced?

RS: The crucial imbalances described earlier cannot be solved through technological wizardry alone. Even with new breakthroughs in science and technology, the social and economic interdependencies involved cannot be dissolved or replaced by some new "magic machine". Even if such a magical machine could be invented, sharing it with the rest of the world cannot be expected, not so long as we maintain existing trade rules and intellectual property models. Worldwide penetration of markets through cross-country supply and value chains means that one dysfunctional market could dramatically impair the wellbeing of countries located on the other side of the globe. From this it can be inferred that we need to acknowledge the many interdependencies that exist today and to help policy makers and societies realize that it is in our best interest to think in "we" terms and not in "us against them" dichotomies.

"Culture" in artistic terms is understood as the means of how we appreciate our life's many expressions or forms of co-existence. They help us see otherness, novelty, and creativity in its many forms at home and abroad. Culture appreciation and production encourages the "creators" and "appreciators" to be open, curious, and willing to engage with other human beings. We need to be open to other people, enjoy the newness of otherness, and reach out to the unknown; this happens when we experience and enjoy art in its manifold manifestations. To nurture a cultural orientation through education and cultural activities, for instance, can greatly help citizens gain insight into the many facets of life and come to terms with the difficulties of today's complexities through deep reflection and a questioning of our current status quo. Art and culture provide opportunities to see today's problems in a different light, helping people get out of a sense of isolation that often reflects a sense of powerlessness and a feeling of hopelessness. Art is about re-imaging the impossible! Cultivating Culture through the arts could help citizens find new and liberating spaces to imagine new horizons!

IWI: If we are to tackle the major social and environmental challenges we face, how should we re-imagine the relationship between ourselves and the environment around us?

RS: We need to re-discover that we are interdependent with our environment and in constant interaction with other people, and that we cannot survive without taking from the environment around us. At the same time, we need to realize that we constitute the environment for others: we either contribute to others through support and care, or we become a hazard to others resulting in rejection or aggression. In the worst case scenario, other people can experience us as a threat. Following basic tenets of existentialist and phenomenological philosophy, as well as Gestalt psychology, it becomes apparent that we need to regain the ability to perceive, to move from perceiving foreground (e.g. the central figure of a portrait) to perceiving the background (e.g. the family or furniture surrounding the figure), and to re-enable ourselves to make contact and move towards others, while at the same time remaining aware that nobody "leaves" the environment for good. Being able to experience the "here and now", so helping us encounter and dialogue with others, would greatly help us identify common ground and a common interest in ensuring the survival of oneself and the survival of others. Art can greatly support this process of becoming aware of otherness and novelty. However, in order to positively integrate these new experiences, we need to create a supportive environment which allows us to experiment and reach out to others. The more this is possible, the easier it will become to reorganize ourselves to jointly face the major social and environmental changes which are awaiting us in the near future.

IWI: James Hansen, the Director of NASA's Goddard Institute for Space Studies, has suggested that "two degrees of warming will lead to an ice free Arctic and sea-level rise in the tens of meters". Although many environmental organizations are lobbying for biodiversity and resource preservation, it seems like it is taking too long to acknowledge the risk of global warming. How could environmental concerns be better integrated into the core mission of the average for-profit company?

RS: As stated in the communiqué of the November 2012 meeting of the UNOSD (United Nations Office for Sustainable Development), Incheon, South Korea, "Sustainable development has been a heavily diagnosed concept over the last twenty years, but we also acknowledge that business-as-usual in the 21st century is not an option. There is urgent need to accelerate implementation and to scale up good practices for sustainable development at all levels". The list of useful studies and suggestions for developing more environmentally friendly business practices is growing longer and longer, as are calls for the creation of more graduate-level academic programs on sustainability management in different parts of the developed world. UN agencies like UNEP, UNCTAD, ILO, and the World Bank, just to name the leading international organizations, continue to organize conferences and write papers on how our economies and enterprise could be made more environmentally sustainable; and still, climate warming is increasing, not decreasing.

In the words of Stefan Schaltegger, one of the pioneering professors of sustainable management, "the main ... discussion about the direction of the corporate sustainability debate [is the] claim that corporate sustainability should go beyond the business case, that it should address issues beyond the environmental aspects and that corporate sustainability should develop its role as a change driver in the economy and society as a whole". With this , the suggestion is made that corporate sustainability should go beyond greening and beyond the company and its stakeholder network. What seems evident is that greening – that is, making a company more sustainable as a micromanagement strategy – is not enough. What is vital is to put the need for greener business practices into the larger societal context, and to find ways of agreeing to more stringent regulatory measures such as incentives (e.g. tax reductions for green investment) and sanctions (punishing companies unwilling to reduce their externalities through fines or, in the worst case, closure).

Click here for a full version of the article with references.

Raymond Saner is Professor Titular in International Relations and International Management at the University of Basle, Switzerland. In 1993 he cofounded CSEND, a Geneva based NGRDO (non-governmental research and development organization), and he is the director of CSEND's Diplomacy Dialogue branch. His research and consulting focuses on international trade and development, conflict studies, and international negotiations at bilateral, plurlateral, and multilateral levels in a number of different fields, including trade (WTO), employment and poverty reduction (ILO, PRSP), trade and development (WTO, UNCTAD, EIF), human and social capital development in the educational sector (GATS/ES/WTO and OECD), and trade, investment and climate change (UNCTAD). Raymond Saner pioneered the field of business diplomacy and contributes to the study of multi-stakeholder diplomacy. He also teaches at Diplomatic Academies and Schools in Europe and North America.

Balancing the Planet with Play

An Interview with Sun Mi Seo

Interviews with Innovators: PlayPlanet is a sustainable travel platform that helps connect local communities around the globe with travelers. Its aim is to help empower grassroots social innovation. What does social innovation mean to you, and how would you define sustainable travel?

Sun Mi Seo: I think that sometimes we tend to confuse innovation with invention. Invention is about creating something new; innovation, on the other hand, is about creating an impact on people's minds and behaviors, which brings about a change of some kind. This is exactly what Play-Planet is trying to achieve, to encourage people to behave differently, to make a change.

Through PlayPlanet we wanted to connect travelers to authentic local cultures and support communities who can offer that access, so improving the experience of traveling and making it more sustainable. At the same time, we also believe that social change can occur through travel, and this is what sustainable travel, for us, is all about. A "sustainable traveler" becomes acquainted with locals and, through this, learns more about local culture and infrastructure, as well as how the local economy can benefit from it. The elements of sustainable travel are, therefore, three-fold: creating benefit for the local community, creating respect for local cultures, and promoting protection of the environment. This is what Play-Planet has taken on as its mission, and this in order to contribute to a global community of travelers who are at the same time participants in a movement for social and environmental change by supporting local communities with alternative incomes.

IWI: Can you briefly describe how the connection between travelers and local communities takes place through the PlayPlanet platform, and what do travelers experience that is so unique?

Our standard host program works by linking "Explorers" – who reveal their interests by choosing online from categories such as learning and education, arts and culture, nature and outdoor – with "Hosts" who can offer their own authentic versions of these experiences. Our hosts are considered as PlayPlanet friends who create, share, and promote our vision. Eddy Susanto, a young freelance guide in the Tanjung Putting National Park, is one such example. Eddy has opened up new doors for travelers, who like himself, want to be independent from mass-produced tourism. We also offer tailor-made trips for institutions or groups to address, for example, projects about local development or environmental protection. A good example of this type of project is our work in Borneo, where travelers learn about orangutans and deforestation. Our network at the moment offers direct contact with local communities in countries such as Cambodia, Indonesia, South Korea, Nepal, Taiwan, Philippines, Thailand, Hong Kong, and Japan.

Through these programs, PlayPlanet offers a unique experience to travelers and provides local communities a way to cultivate economic value through micro-entrepreneurship. In this way, traveling can make an impact not only on travelers but also on local communities around the planet.

IWI: Low-cost airlines have made travel more affordable and travelers more " fun-oriented". How does PlayPlanet keep the balance between play (fun) and planet (sustainability)?

SMS: We are still working to find and maintain the balance between play and planet, or between fun and sustainability. It's clear that having a fun-oriented approach is very important to the industry. We have experimented with different approaches in the past, and some weren't a success because they focused too much on social and environmental issues;

some clients felt that we weren't giving them enough opportunities to have fun. So, focusing too much on issues around sustainability isn't going to help us attract travelers.

We want to be understood as a platform that offers opportunities by bridging possibilities. Anyone who has local roots and a passion for local culture can become part of PlayPlanet by simply uploading information to our platform; this can be done independent of educational or financial background. This is how we are trying to create a new approach to traveling while redefining both our business model and the role of the consumer. Currently, we are re-designing the PlayPlanet concept with a focus on accentuating the fun aspects of traveling. We want to promote what we offer as a playground platform, where travelers engage with locals and vice versa for fun.

IWI: How do you manage the PlayPlanet network, and what are the innovative or nontraditional business practices involved in your approach?

SMS: Although we are not a traditional business, we do exhibit some similar traits. For example, we have contracts between PlayPlanet and local hosts for projects and activities. We base our business practices, however, not only on profits and revenues but also on social trust, a trust that is rooted in the main structure of our network. That is why we work hard to understand each local context in depth; based on the findings of our research, we contact potential partners and then try to build a sustainable collaboration with them that eventually provides some form of income for us as well as for the locals. As such, PlayPlanet is a people-content platform where personal stories really matter, and where standard tourist attractions matter less. This is why we advocate trust-based collaborations, identifying so called "Local Heroes" who are the key to changing people's traveling behaviors. Overall, you could say our business model is to create social impact that encourages more people to understand and participate in sustainable travel, to support the idea of travel as a shared economy.

Our future goal is to give more management freedom to our local partners so that we don't have to be over-involved in the collaboration.
Through these efforts, PlayPlanet is becoming more and more finically
sustainable as a company. We started with only 100 dollars, and have
never received any public financial support. Although the income from
PlayPlanet cannot be compared with the income of a more traditional,
similar-sized business, we are still expanding and bringing in new people
to our team. We are positive that in the future our salaries will increase.

*IWI: Thinking back to the beginning of this story, what inspired you to ini-
tially set up PlayPlanet?*

SMS: As a student, I used to travel a lot on a very low budget, mostly
backpacking and staying with locals most of the time. For me, traveling
always offered such unique and special experiences. Getting to know foreign cultures, languages, and local cuisine was an adventure. However,
for most of my friends from the Philippines or Cambodia, travel has a
completely different meaning. Travelers are considered to be rich foreign-
ers or curious consumers. This helped me to realize that there is a huge
gap between travelers and local communities in terms of how they feel
about travel and how they could benefit from it.

At the beginning of my career, I was working in the field of social enterprise and tourism. I had to design a sustainable tourism program for a
tourism agency, and felt very uneasy about doing so. Although there are
many companies in South Korea working on commissioned projects to
create sustainable travel experiences, in Europe it's more about raising
awareness and empowering local communities, something much closer
to what I wanted to do. My idea was to contribute in a sustainable way
to the life of my local friends.

This was essentially how PlayPlanet was born. Later on, whilst traveling,
I met a couple from Thailand and presented my idea of a sustainable
travel platform to them. They were IT specialists, liked the idea and offered to develop the project with me. We still see each other two or three

times a year in Thailand or in South Korea, and continue to work together on ways of improving PlayPlanet.

IWI: The work of PlayPlanet comes under the broader umbrella of the creative economy. Do you feel that this sector is a smart choice for the future of South Korean enterprises and start-ups?

SMS: The South Korean government is really pushing the **creative economy** at the moment, but many people are overusing the word creativity in an inappropriate way. Yes, young people are becoming empowered to create new businesses that have a creative flair, but they aren't always creative in how they open up opportunities for building new, sustainable business models. There are more educational events, seminars, and lectures available than ever before where people can meet, network, and learn about alternative economic approaches, but people have a limited understanding of what the creative economy is, and how society can benefit from it. For example, in South Korea the tourism industry is really dominated by huge companies who monopolize the tourism market. Those companies are profit-oriented and don't really contribute to society at all. PlayPlanet was conceived as an alternative to this corporate landscape, to drive a creative change in the existing tourism sector. We are definitely an example of an enterprise that contributes to the South Korean creative economy through exploring new types of sustainable business model that can also have positive social and environmental impact.

IWI: What are the biggest challenges for a female social entrepreneur in South Korea or worldwide?

SMS: Sometimes gender really counts when it comes to business. While setting up PlayPlanet, a number of potential investors and local partners wouldn't take me seriously because of their traditional views on the role of women in society. Unfortunately, South Korean society still puts men first, and this is because many women marry and stop working early on. I think some investors might just have assumed that female PlayPlanet employees would soon get married, have children, and then stop working. So, a lot of my effort has been expended in trying to convince people, and to show them how much I care about the social aspects of our business. In most cases they then change their attitude. Still, as a woman, I have often felt hurt, and sometimes so much so that I have become really frustrated. This has begun to change recently, however, and these days everyone around me thinks that it's really cool that I am a young woman who is a CEO. It shows that opinions can be changed.

I think that being a woman has allowed me to create a welcoming and distinctive working environment through creating a soft, detailed, and flexible working culture. I think focusing on approaches to business that are based on sustainable practices can actually help get more women involved in the business world.

Sun Mi Seo is the founder of PlayPlanet, an online platform that helps connect local communities around the globe with travelers. In this way, PlayPlanet hope to lead sustainable travel by benefiting local economies, preserving local culture, and protecting the environment. She was a founding member of Travelers' MAP (Travelers Make An Amazing Planet) established in 2009, which was the first social enterprise for sustainable tourism in South Korea. In particular, she's focusing on how technology/ICT can help to build an ecosystem where locals and travelers can co-exist, collaborate, create impact, and develop sustainable models of tourism.

Learning By Gaming

An Interview with Peter Lee

Interviews with Innovators: You are an accomplished game designer with an interest in developing innovative game-based learning approaches for children and adults alike. You grew up in South Korea but migrated to the US, a move that has proven critical in your development as an entrepreneur, artist, and game designer. Are these separate roles for you – with separate hats – or do they rather form a well-tailored three-piece suit?

Peter Lee: Well, migrating to New York was important to me. I met a lot of people who helped me forge my identity, and who contributed to me getting off to a good start in what I'm doing. I met Erik Zimmerman and also Katie Salen, for example, who were instrumental in helping me obtain several grants to develop new ideas. Together, we started the Institute of Play and realized projects like "Quest for Learning". Looking back, I started as a game designer but slowly shifted to entrepreneurial work as my interests changed from video games to game culture and education. So, I guess it's more like a three-piece suit!

IWI: In the past you have spoken about a number of disasters that have deeply influenced you, such as the TWA Flight 800 disaster in 1996, the boom and bust cycle of the 2001 Dot.com Bubble, and the 9/11 terrorist attacks on the United States. How did these influence you in terms of the way you create art or do business?

PL: I have to say that those disastrous moments in history really hit me. When I was a student, I strongly believed that I should always be prepared, that I should manage my life so that when such a day came I would not be negatively surprised. I realized after the 9/11 incidents that tragedies like this come unexpectedly; we cannot really foresee them, nor can we manage them in advance. There is simply no way that we can have full control over what happens in our lives. Reflecting on this, I became softer and more flexible in my approach to life, and I stopped worrying so much about the future. This element of uncertainty, or "lack of permanence", is also really important to those people interested in gaming. It is an element that keeps people engaged in game-play and increases their motivation to win.

IWI: You co-founded the video game company GameLab in 2000, and then helped start a non-profit organization called the "Institute of Play" in 2006. What is the relationship between these two different types of venture?

PL: GameLab was a game development studio that I founded with game designer Eric Zimmerman. Our aim was to foster a culture of design research and implement rigorous creative processes into the daily operation of the company. The GameLab encouraged the creation of experimental and innovative games, which helped define the independent game studios of today – and this many years before "indie games" became a buzzword. Our office was in downtown Manhattan, and we released around thirty-four video games for multiple platforms between 2000 and 2009. These were published by companies like LEGO, HBO, PlayFirst, etc... One of the socially orientated games we made, for example, was about poverty in Haiti created with Global Kids and a class of high school students.

IWI: How was GameLab, and its interest in socially orientated gaming, received in New York?

PL: Our approach had a strong impact on independent game culture, especially in New York City. In New York, GameLab was considered to be an innovative game-designer cluster that could produce games which were not only fun-oriented but also had a strong, inbuilt social element. People really liked them, and – because my partner Eric Zimmerman had great networking skills – we were able to make lots of important connections such as with the MacArthur Foundation and the Bill & Melinda

Gates Foundation. As a result, when the idea of creating an educational gaming platform came up, we were able to find support and partners for it quickly.

So, in 2007, GameLab spun off into the non-profit "Institute of Play" to promote game design and play as educational tools for students. The games produced by the Institute of Play were basically focused on creating learning experiences rooted in the principles of game design, experiences that simulate real-world problems. Through the institute we were also supporting and training teachers to make learning irresistible for students through gaming methods.

IWI: Can you give us an example of a project that was particularly important for the Institute of Play?

PL: Our first initiative was "Quest to Learn". Quest to Learn is a public school within the New York City Department of Education system, and it came about as the result of a partnership between the Institute of Play and the New Visions for Public Schools program. The school runs an innovative learning model based on a "game-like learning" approach. What's really important is that the school's curriculum is co-designed by teachers and game designers to create both a long lasting partnership and a visible impact of game design on the educational system.

IWI: You recently moved back to South Korea, where you set up a company – called Nolgong – that focuses on game play for adults and children. Can you tell us about your experience of working with schools in South Korea?

PL: I arrived in South Korea with a strong mission to make learning a natural and playful experience. This is how Nolgong was born with the aim of introducing game-based learning approaches into Korean Schools. The Korean education system is very rigid, with students often overwhelmed by the quantity of information that they have to memorize from books. There is also little interaction or collaboration in the process of learning. While observing children playing in schools, we have noticed that they are only taught to be well behaved, rather than encouraged to experiment, to play, and to deal with the emotions they may be feeling. In the game-based learning model, players need to feel that they can be open about the choices they make and show their feelings to the people around them. This increases their social skills and can create a collaborative achievement that impacts both the individual and the group. So, what we want to achieve within the Korean educational model – in order to break the rigidity of learning – is to encourage children and students to engage with their emotions and develop a sense of social interaction. We hope to achieve this through new forms of educational and experimental gaming.

IWI: Do the approaches you've developed at Nolgong have a wider application within society, such as in the commercial world?

PL: Yes, absolutely. We have, for example, created projects for major Korean Companies such as Hyundai and Samsung. Corporations are very keen to explore how the element of play can be used to improve the social skills of their employees and increase a sense of cohesion within the company. In developing these new game-based training for corporate partners, we have also experienced a lot of freedom in how we create different types of learning experience suited to different tasks. This has opened up new possibilities for working and interacting with a wider variety of people, from young students through to senior citizens. Here at Nolgong we enjoy working with different types of client – whether for-profit or not-for-profit organizations – because it is important for us to keep the company financially sustainable and adaptable in the future. We don't want to rely only on public funding.

IWI: One of your concepts is that play cannot be designed, only rules can; play emerges when players interact with those rules. Can you explain this idea further using an example from your own work?

PL: A game is an outcome of rules. All games – digital as well as analog – begin with a clearly defined identity, one which helps engage players and aid them in developing a sense of the main characters. Also important to game design is a game's vision and guidelines, as well as how they contribute to a player's gaming experience. So, making rules is important for the structure of the game, its identity, and the system of emotions it operates on. Regardless of digital or analog game design, our rules aim to encourage the players to act as engaged collaborators, and help them get immersed into the game to a point where new feelings and emotions are triggered.

In the realization of the project Being Faust (a project produced in partnership with the Goethe Institute in Seoul), we tried to give the players an opportunity to feel and experience the core message of this classic work in a new way. By integrating the element of mobile gaming into the project, we were able to create a real-world gaming experience, essentially a physical game enriched with online and social media elements. This not only created a more personal experience of a work exploring where "values and ideals are up for sale", but gave us very individualized feedback in return, feedback which will help to us in the development of future projects. The Goethe Institute was keen on a collaboration that would result in new interactive ways of learning. I personally think the project worked out perfectly; we have certainly received a lot of positive feedback from visitors to the Goethe House in Seoul.

IWI: It seems that one way of implementing change is to break the rules, the benefits of which may only be seen by those who come afterwards. Do you think making new rules, rather than breaking old ones, could be a less risky approach, but one that still leads to positive change?

PL: Making new rules can also involve breaking old ones. But, it's my belief that it is more important to focus on creating change rather than on breaking rules per se. I truly admire those people who can actively create new rules – and thus drive change – without completely breaking apart what has come before. I see, for example, the Occupy Movement not as about breaking the rules, but rather about forcing change and pointing out where change is needed. With our approach to using gaming in education, we hope to nurture students in making use of rules in their own way, rather than have them follow rules in a ridged manner or break them simply because they can.

Peter Seung Taek Lee is a game designer with an interest in creating new forms of play that can redefine art, culture and society. Following a Bachelor of Fine Arts degree in Computer Arts at New York University, Peter went on to obtain a Master of Professional Studies from NYU's Interactive Telecommunications Program. Peter co-founded GameLab in 2000, a successful video game company in New York. Later, in 2006, Peter helped start a nonprofit organization called the Institute of Play. One of the key projects masterminded by the institute has been Quest to Learn, a public school in New York with a gaming-based curriculum that aims to make learning a natural function of play. Now back in South Korea, Peter is co-founder of Nolgong, a company that aims to introduce game-based learning approaches into Korean Schools. Since 2011, Nolgong has worked on making learning a more playful experience for over 20,000 children and adults.

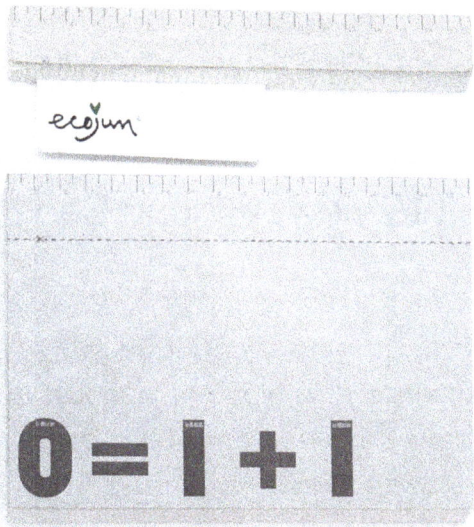

ecojun

0 = I + I

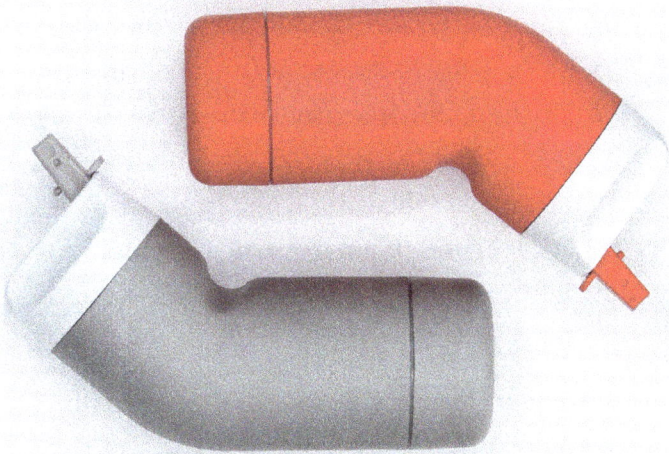

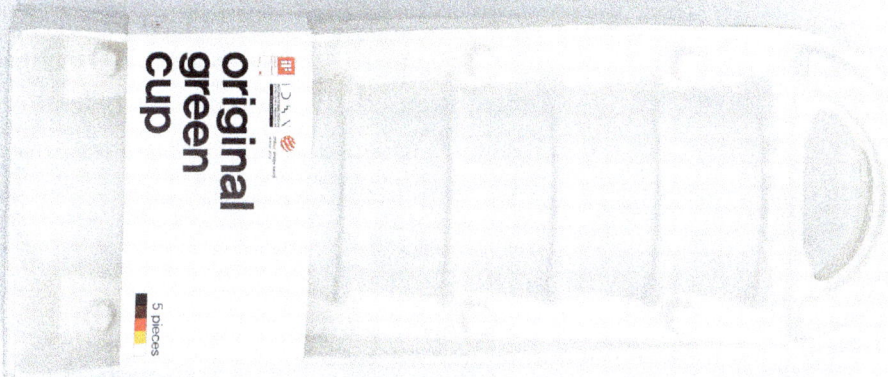

original
green
cup

5 pieces

Benchmarking Eco-Friendly Design

An Interview with Jun Seo Lee

Interviews with Innovators: You are the founder and CEO of Ecojun, a company that takes the wellbeing of both humans and nature into consideration when designing products. Can you tell us how your interest in sustainable product design began, and how this has led to the creation of a successful company that embraces such principles?

Jun Seo Lee: During my college years, I originally planned to become an advertisement designer. I successfully built up my portfolio and participated in many contests, winning around 25 prizes. One day, however, I met Professor Hoseob Yoon, the founder of Green Design Department in the graduate school of Design, Kookmin University. He introduced me to the concept of "green design," which then become my main focus after 2005. Around this time, the term "green design" was poorly understood in Korea, so I faced many difficulties in following this path. Still, I strongly believed in the ideas I had learned from Pf. Yoon, which included a focus on philosophy, ethics, and ecosystems rather than design skills.

I believe in the values of green design and both the role and responsibility of designers in protecting our environment. Ecojun is a company the has come out of this way of thinking. It was established to introduce green design into Korea and share "green-compassion" with consumers.

IWI: Scientists are looking for solutions to environmental issues through the development of new technologies and policy change. Environmental problems are, however, also being tackled through art and design practices. What message does Ecojun wish to send to its consumers about the importance of protecting the environment?

JSL: When it comes to environmental issues, the role of the scientist is to forecast future problems and develop potential solutions. The role of government, on the other hand, is to legislate on those issues currently faced and to raise public awareness. The role of the designer, I think, is to interpret highly technical scientific ideas and government policy in order to generate implementable solutions that can work with real people in the real world.

Ecojun, with these beliefs, aims at providing practical solutions rather than trying to force behavioral change. Our green design products, therefore, target not only green-sumers but also the general consumer who might be interested more in product convenience than environmental protection. When green design products can attract consumers through good design – rather than just through their design philosophy – an "eco-lifestyle" can be more widely, and effectively, promoted. In this way, Ecojun is tackling environmental problems through creating a message of ethical consumption, as well differentiating its products within the marketplace.

IWI: You have been awarded with some of the most prestigious design awards, such as IDEA, IF, Red Dot, and UNESCO creative Design. What measures of success and impact do you adhere to in product design?

JSL: Normally, the success of a product is defined by product sales. Here at Ecojun, we define success according to the level of CO_2 emissions during production. Understanding all the stages of a product's life cycle (through carrying out a Life Cycle Assessment) and minimizing CO_2 emissions through design are difficult tasks. Such a pursuit, however, can also be a creative process – even the happiest moment of any product design process. When consumers understand these ideas and respond to the challenges we are tackling, we designers feel that we have succeeded. Wouldn't we be considered successful if the work of designers and the behavior of consumers could be linked to generate a virtuous cycle of green design and eco-lifestyles?

Consumers today are surrounded by adverts; they don't turn their attention to things that don't interest them. However, if you can capture someone's interest, they will pour time and effort into finding out more. In these circumstances, companies like Ecojun – whose goal is to send a message of social, economic, and environmental responsibility to society – should focus on arousing interest amongst consumers through good design. Good design draws attention to products and the messages that lie behind them.

IWI: Today, many people are concerned for the environment and its future, leading them to follow principles of green consumerism and to purchase products based on ecological sustainability. What do you think motivates people to make ethical choices even in times of economic austerity?

JSL: Choosing to strive for ecological sustainability is no longer an option. Economic and environmental pressures largely contradict each other: Greater economic activity means greater production and consumption, resulting in the increased generation of waste and pollution. We can show these links statistically. For instance, when the USA entered into recession and its economic growth dropped 5%, the percentage of annual waste dropped by 14.5%.

It is important that we replace common products with eco-friendly ones. An increase in demand for eco-friendly products will result in price reduction, making them available to even more consumers. However, making ethical choices for the environment does not necessarily mean buying eco-friendly products. I usually say in my public lectures: "If you already have cups or bottles in your home, then use them. Don't buy Ecojun products simply because it's trendy to do so. This is our message: Add value to the goods you already own. This is an important part of the path towards ecological sustainability."

IWI: Good design excites people, can attract interest to a product, and can even drive behavioral change. Ecojun's idea is that when people begin to use design not only as a marketing tool but as an environmental tool, they will raise awareness around environmental issues. With this in mind, and knowing that your products have served as benchmarks in eco-friendly design, can you tell us about a few of your products and the impact they may have had on the public mindset?

JSL: Any number of design companies can pursue the creation of desirable products. What differentiates Ecojun from its competitors is that we consider more deeply the message we wish to send, the audience we wish to reach, and the means by which that message is communicated. Green-sumers who buy eco-friendly products have had, for a long time, little choice in this market. They have had to sacrifice their desire to purchase desirable and attractive products in order to pursue their eco-oriented beliefs. I think we designers should offer much more choice to those green-sumers.

Unfortunately, there are not many designers in South Korea who are interested in green design. However, as the eco-friendly mindset becomes more prevalent, I hope that more designers will turn to green consumerism. To these ends, designers should try to understand why green-sumers practice green consumption and explore how they can change existing prejudices against eco-friendly goods. For example, I have found that many green-summers buy eco-friendly products out of consideration for their own health. So, Ecojun produces goods which target this need amongst consumers, whilst, at the same time, communicating a strong message around environmental concerns.

IWI: The over-use of the paper coffee cup has contributed to the widespread deforestation that has severely impacted our planet's capacity to remove carbon dioxide from the atmosphere. Can you tell us about Ecojun's "Green Cup" and how this might lead to the demise of the paper coffee cup?

JSL: The Ecojun Green Cup has been developed to reduce the use of paper cups in our daily life. Ecojun has sold 150 thousand Green Cups since 2011. If consumers used a Green Cup 10 times, this would be the equivalent of 1.5 million less paper cups used. Since the production of each paper cup generates 12 to 15 grams of CO_2, 18 to 22.5 metric tons of CO_2 would be saved.

A key characteristic of the Green Cup is its V-groove which holds the tea bag string safely to preventing the bag from falling into the cup. The V-groove is a very simple idea, but it appeals to users. This means they'll use it more often, so reducing their consumption of normal paper cups. Furthermore, the Green Cup is eco-friendly even after its disposal. As it's made with degradable cornstarch, which doesn't emit any harmful chemicals during degradation, the cup can be safely thrown away.

IWI: Ecojun's "Public Capsule" is a portable drink capsule made from biodegradable plastic material. With each sale of the capsule, a contribution is made to towards Malaria prevention pills for children in Africa. Can you tell us how campaigns for social responsibility have impacted the company's design process and mission?

JSL: Ecojun doesn't only focus on environmental values; we look at other areas of ethical consideration as well. We try our best to be a driver for positive change in this world and to take our social responsibility seriously. The Public Capsule is the result of this endeavor. Honestly, it is a real challenge for social enterprises to be both economically profitable and socially responsible. Ecojun sometimes fails to convert meaningful ideas into actual products too!

Nonetheless, we are pursuing these values because it enriches us. Brainstorming for better, creative, and more socially valuable products helps us to overcome many of the struggles and hardships we face. We launched Public Capsule in order to contribute towards Malaria prevention in Africa. We have also launched a new project to support the provision of water sources in Ethiopia. It is the energy we get from such projects that drives us forward.

IWI: Another product from Ecojun is called "1+1=0." This is a notebook made from recycled paper that down-cycles into a business card holder.

More than simply a notebook, it is an eco-friendly design statement. In which product areas do you feel issues around green innovation can be made?

JSL: The environmental problems we face today are neither simple nor one-sided; rather, they are very complicated and multifaceted. For this reason, Ecojun is trying to expand its product range to target more diverse groups of people.

The notebook "1+1=0" is a concept design, developed to inspire eco-friendly ideas amongst fellow designers. The Earth consists of diverse materials and life forms. Water, air, soil, animals, plants, and mankind cannot survive without each other: balance is the key for successful coexistence. This principle applies to industry also. It cannot be said that only certain product area need be reformed while others can be left alone. Inventors, producers, and designers in every industry should consider the development of eco-friendly and sustainable products not as a luxury, but as a necessity. If they don't, we are gradually going to erode the beauty of our living planet.

Still, if I have to pick one or two product areas where green innovation and statements of sustainability should be made, I would choose the aircraft industry and mobile telecommunications. An airplane, in a single flight, exhausts the same amount of CO_2 as 1800 cars. There is a need for green innovation in this sector to reduce greenhouse gas emissions. Similarly, mobile phone companies release new models of their phones every one or two years, resulting in increased waste from discarded devices and the continuing exhaustion of natural resources (as well as creating other environmental and social problems). Greater attention to issues of sustainability in the mobile phone industry are required.

IWI: Finally, the Ancient First Nation Proverb says: "We do not inherit the Earth from our Ancestors, we borrow it from our Children". How do you think we can cultivate a culture of eco-literacy to teach younger genera-

185

tions to be "smart with nature", and do you think the arts can play an important role here?

JSL: I believe art can be adopted as a powerful means to share and communicate the value of our natural ecosystems with younger generations. For example, artists can create performances revealing the ways in which we are polluting our environment, so raising awareness of the issues we face and mankind's role in them.

The many environmental problems we face today are the result of our own activities. We cannot expect our own children to solve them alone; we have to work with them. Rather than throwing around facts and statistics, we need to educate our children to develop a sensitivity towards the environment. This has to start from very humble beginnings. I cannot simply declare "Save the Earth!" I must teach my son how to take care of our own neighborhood first. Therefore, I sweep the block in front of my house with my son. Step by step, I can convey to him important messages about green innovation and environmental sustainability. I believe the most valuable thing we can pass on to our sons, daughters, grandsons, and granddaughters is a clean, beautiful, and happy Earth.

Ecojun is a design company that develops and produces green design products based on environmental and ecological thinking. Substituting traditional materials with eco-friendly ones in the manufacturing process, Ecojun is pursuing the design of universal products that consider both consumer convenience and green design credentials. The company's management philosophy is to offer customers with high quality product designs that can also benefit the environment for future generations.

From Collisions to Collaborations

An Afterword by Jeungmin Noe

As a curator with many years of experience in promoting art and technology, it's fair to say that science – its materials and practices – is still a topic that lies outside of normal curatorial interests in South Korea. However, now that more artists and scientists are becoming interested in cross-over activities, I hope the value of museums in promoting and evaluating such acts of convergence will be recognized. Bringing research findings from The Human Brain Project, or participatory social housing projects, into the museum has been a new and exciting experience for us. Such activities not only broaden the museum's horizons, but also open up our galleries to a more diverse audience to the benefit of the community at large.

The most exciting aspect of hosting the DAW Festival was the opportunity to re-examine the relationship between the arts and sciences. As the diversity of connections between different disciplines has grown, the full richness of possibilities for convergence has started to emerge. With the advent of digital technology, this shared canvas on which both artists and scientists can work continues to expand, offering new ways of seeing, feeling and experiencing the world around us. The exhibition presented work that delighted in crossing disciplinary boundaries, work that tests our understanding of what disciplinary activities might be. In this way, perhaps, the exhibition points us towards a new and exciting world; scientific knowledge assisting the creation of a new era in the arts; artistic practices assisting the creation of a new era in science.

The power of collaboration across disciplines lies in how people can create new and unexpected paths. This book explores some of these amazing moments, and offers ideas for others to take even further. It is exciting to imagine what will happen next.

Glossary

Augmented Reality (AR) describes an augmentation of real-world experience through the overlay of digital media (such as sound, video, graphics, or GPS data) in real-time. Conventionally implemented through handheld mobile devices, AR allows digital content to become responsive to the location and behavior of the user. Acommon application of AR is to introduce a digital artifact (such as a model of a building) into the screenimage captured through a device's camera functionality; geo-located, the model appears to be part of the physical world, remaining in its 'physical' location as the user moves around with the device.

Bioart describes a form of artistic practice in which artists manipulate, or modify, living materials such as tissue cultures or bacterial organisms. Using scientific methods in a laboratory, gallery, or artist studiosetting, this living materialbecomes the foundation for new artistic work. Bioart practices might be inspired by, or used to comment on, specific issues in biomedical science. As such, it is rapidly becoming a significant area of work in the field of aesthetics, with far reaching implications for what we understand the nature and role of art to be. By working with living materials, Bioart practices always have ethical implications.

Biohacking refers to experimentation with biological systems outside of an institutional setting by amateurs who identify with a "hacker" ethic, one founded on hands-onengagement, knowledge sharing, collaboration, and community. Biohacking activities might include technological interventions into the body, the gathering and use of personal bio-data, and "do-it-your-self" approaches to microbiology and genetic engineering. Often taking place in small and informal labs (hackerspaces), the movement is strongly allied with citizen science activities, those working to make the tools and resources necessary for conducting basic science experiments available to all.

Citizen Science refers to the inclusion of the general public and science enthusiasts in science projects, with the degree of participationshaped by the institutions, protocols, materials, and data involved. Simple forms of participation include observation work and data collection, conducting simple cognitive tasks, undertaking crowd-funding work, and making donations oftime or computer power. More complex types of citizen science activities include corroborating research, initiating new research agendas, creating informal institutions (e.g. hackerspaces), and contributing to the public understanding of science and science policy-making.

Corporate Social Responsibility (CSR) describes a form of corporate policy that lays out the activities of business entities-deemed to contribute to society at large. Through CSR policy, businesses may strive to promote the ethical and equal treatment of employees, donate to national and local charities, reduce their carbon footprint, and so on. CSR policy differs around the world, coming to reflect unique constellations of social and cultural values. While some businesses exceed expectations in their contribution to broader societal goals, others may feel either unable, or uncompelled, to pursue them. There is currently a debate amongst business leaders as to whether businesses have the democratic legitimacy to take on CSR roles.

Creative Economy refers to those economic activities involving the use of creative talent for commercial purposes. Although centerd on the creative and cultural industries, the notion is a broad one, recognizing creative occupations irrespective of the industry in which they are placed. Many have come to argue that the creative economy not only generates economic growth and employment, but is fast becoming a key driver of innovation more generally: creativity as a primary resource for manufacturing attention, complexity, identity and adaptation.

Digital Heritage describes digital information derived from sites of social or cultural value. Practices involved include the two- and three-dimensional modeling of artifacts; the use of augmented and virtual reality tools to contextualize digital resources; the use of data mining and information extraction to analyze cultural heritage texts; the development of tools for preserving, disseminating, and remediating digital data to different audiences within heritage contexts. The field is currently experiencing rapid growth as it explores new trajectories for cross-disciplinary research and the potential for fruitful collaborations between academia and the creative and cultural industries.

Microenterprise is a term for enterprises that employ fewer than ten people and whose annual turnover isless than two million Euros. In Europe, 99% of all businesses are micro, small or medium size enterprises, with nine out of ten of these categorizedas micro. Worldwide, the microenterprise sector is a hotbed for innovation and entrepreneurial activity, contributing significantly to economic growth, social stability, equity, and job creation. Entrepreneurial activity in the sector is extremely varied, reflecting both the desire to exploit new opportunities in a given market, but also the need to generate basic forms of income.

Neuroaesthetics is concerned with the neural underpinnings of the aesthetic experience of beauty, particularly, although not exclusively, in relation to the visual arts. The term aesthetics used here broadly encompasses the perception, production, and response of individuals to artistic work.

The field draws on a variety of disciplines such as psychology, evolutionary biology, philosophy, and neuroanatomy. A relatively new and controversial field, neuroaesthetics is searching to reach higher levels of reproducibility in its publications, handle criticism of its reductionist methodologies, and refine its conceptual and practical approaches to probing the aesthetic experience.

Open Hardware refers to any type of electronic hardware built from designs that have been made available for public use at no charge. Designs might include schematic diagrams, construction plans, parts lists, source codes, and documentation. In contrast to closed hardware practices in which patent law is used to inhibit the reproduction of the object, the goal of open hardware is to make the object as easy to reproduce and redesign as possible. Open hardware does not lend itself to the open source paradigm in the same way as software; although documentation and source files can be made available free of change, most open hardware projects cannot afford to offer physical components gratis.

QR Code (abbreviated from Quick Response Code) describes a type of two-dimensional, machine-readable barcode. Unlike a traditional one-dimensional barcode, which encodes information through the different thickness and spacing of parallel lines along a single axis, a two dimensional barcode uses both horizontal and vertical axes, so considerably improving information storage. Taking on the appearance of a patterned chequerboard, the physical QR code is scanned with a smartphone application to activate targeted digital media (such as videos, texts orimages). In enabling a layer of digital content to be added to objects, QR codes are now finding use in product marketing, artistic production, and museum practices.

About the Editors

Arthur Clay has designed and implemented trans-disciplinary events focusing on creatively connecting art, science, and technology within a diverse cultural contexts in many parts of the world. As co founder and artistic director of the Digital Art Weeks his activities include developing and platforming projects that pioneer new technologies include working with renowned institutes such as EPFL, ETH Zurich (Switzerland), University of the Arts Zurich, Xian Academy of the Arts, Xian (China), Sogang University (Korea), Seoul National University (Korea), and for various private and government agencies and institutions including Human Brain Project, SAST, Create Center at NUS, A2Star (Singapore), and more. He has been supported by governmental agencies, art councils, private foundations, and industry partners such as Presence Switzerland, Pro Helvetia, Swissnex China, Swissnex Singapore, Japan Foundation, the Canada Council, LG Electronics. and many others He is also renowned as a versatile artist working in diverse genre and has been awarded prizes for performance art, media art, music theater, and composition. As educator, he has taught over the past ten years at a number of high ranking institutes in diverse countries in Europe, North America and Asia. At present, he is Guest Professor at Sogang University of Seoul (South Korea) and directs the Virtuale Switzerland, a festival for virtual arts and urban gaming.

Monika Rut is cultural researcher interested in arts and culture in future cities and hybrid models for audience participation. She holds Masters in

Arts Management from University of Bologna (Italy) and KMM Institute of Media and Culture Management in Hamburg (Germany). Since completing her studies, she has worked with a number of international organizations and universities including Kulturprojekte Berlin GmbH (Germany), European Culture Foundation, Academy of Visual Arts and Design (ABADIR), ETH Zurich (Switzerland) and Digital Art Weeks International. She was the project manager for the DAW Innovation Forum Seoul 2014 and curated the content of the conference. She is also a co-founder of Virtuale Switzerland.

Timothy Senior is a scholar and artist, currently serving as a Knowledge Exchange Researcher for the Arts and Humanities Research Council, UK. His work asks how contemporary forms of practice in the arts, sciences and humanities might be opened up to new collaborative influences. Following his D.Phil. in Systems Neuroscience (University of Oxford 2008), he has explored these issues through an artist residency at Duke University (USA) and visiting lectureships at Jacobs University Bremen (Germany) spanning the arts, neuroscience, digital humanities, and the social and political sciences. In 2012 he was awarded a Junior Fellowship at the Hanse Institute for Advanced Study in Germany, conducting research on the emergence of innovative performance-based methods at the intersection between contemporary artistic and scientific practices.

Jeungmin Noe co-curated the "Hybrid Highlights – Switzerland and Korea" and Innovations Fora at the Museum of Art, Seoul National University. Seoul National University (Korea) is her alma mater. She received her master's degree from Cornell University (USA), going on to complete a Ph.D. at the University of Iowa (USA) with a dissertation focused on the museum experience. Her interest in the development of new pathways for contemporary society has led her to create such forward-looking exhibitions as "Data Curation" and "Design Futurology". She is currently working towards the construction of a new museum for children, where she will take on the directorship.

Se-Jin Stellar Park (Editorial Assistant) is currently a student of Philosophy at Sogang University in Seoul (South Korea).She participated in the Digital Art Weeks 2014 Seoul as a junior project manager for the Innovation Forum Conferences. She also served as a translator during the Fora and editorial assistant for the "Interviews with Innovators" Book.

Digital Art Weeks International: Founded at ETH Zurich (Switzerland) in 2005, DAW International has the goal of driving interdisciplinary initiatives that bridge the arts and sciences within diverse cultural contexts. Under the motto "art and science creatively connected," the festival program presents diverse perspectives on innovation in art, science and technology through authoritative voices from around the world. Consisting of conferences, exhibitions, workshops and performances, the DAW program offers insight into current research in art and technology as well as illustrating the synergies that result, making artists aware of current trends in technology, and scientists aware of the possibilities for applying technology in the arts.

www.ingramcontent.com/pod-product-compliance
Lightning Source LLC
Chambersburg PA
CBHW081503200326
41518CB00015B/2362

* 9 7 8 8 7 9 3 3 7 9 1 2 1 *